Calculus II Workbook
100 Exam Problems With Full Solutions Covering

Integration Applications
Integration Techniques
Introduction to Differential Equations
Sequences and Series

D1567351

N. Rimmer

ISBN 978-1-7324159-0-4

Table of Contents

PREFACE

This is a collection of my Calculus II midterm exam problems. I (using methods taught during lecture) wrote the solutions. There may be an easier way to solve some of the problems, as with any question, there are multiple ways to approach the problem. If you happen to find a mistake please don't hesitate to contact me (nrimmer@calccoach.com) to point it out. This workbook is meant for any person studying Calculus II which is normally a second semester Calculus course. This is my second workbook of this type. In 2017 I published my Calculus III Workbook, you can find it here: **https://tinyurl.com/ya2jrrdh**. It is my hope that these workbooks will aid in learning the material. The workbook together with a good set of notes and lecture videos serve as a great education package. I am working to post lecture notes and lecture videos on my website **calccoach.com**. Be on the lookout for more study aids.

This book is dedicated to my family, friends and students.

That's a long list, here are some highlights:

Immanuel, Sariah, Mom, Dad, Ricky, Jessica, Reggie, Ray, Marc, Taniel, Shané, Anushka, AC and all my students past, present and future.

Section 1: Integration Applications

Volume

(*a*) Cross Sections:

Slice $\perp$ to x – axis $\Rightarrow$ do the integral in x $\qquad$ Slice $\perp$ to y – axis $\Rightarrow$ do the integral in y

Get a good sketch of the region, find the bounds of the region

> For example: If integrating in x, finding the lowest possible x in your region, this is the lower bound of integration. Do the same to find the upper bound. This is usually done with algebra (setting the functions equal to each other).

Draw a line in the region to represent the slice, label the slice s

Have the formulas for all of the areas of simple shapes

Write a formula for s in terms of the integration variable, call it $A(x)$ or $A(y)$

$$\text{Volume} = \int_a^b A(x)\,dx \quad \text{or} \quad \text{Volume} = \int_a^b A(y)\,dy$$

Volume

(*b*) Disk/Washer vs. Shells

Get a good sketch of the region, make the rotation axis dashed, make 2 copies of the sketch

On copy 1 draw a rectangle in the region that is $\perp$ to the rotation axis

> This copy is for setting up disk or washer

> > If the rectangle is flush up against the rotation axis all throughout the region $\Rightarrow$ **Disk**

> > If there is a gap between rectangle and the rotation axis all throughout the region $\Rightarrow$ **Washer**

On copy 2 draw a rectangle in the region that is $\parallel$ to the rotation axis.

> This copy is for setting up shells

Make a note on each copy indicating the integration variable, it is determined by the orientation of the rectangle. $\left. \begin{array}{c} \\ \end{array} \right| \Rightarrow$ integrate in x

$\rule{1cm}{2pt} \rightarrow$ integrate in y

Helpful thoughts on choosing a method:

> (*i*) When in doubt, choose the x route

> (*ii*) Be on the lookout for the rectangle having an upper or lower bound that switches curves, this will require 2 integrals. Avoid this if possible.

> (*iii*) If the upper and lower bound are ever on the same curve, it is most likely undoable

Volume

(b) Disk/Washer vs. Shells

Once you decide on a method, know how to find what the method requires :

Disk Method :

radius is the height of the rectangle

bounds found from moving the rectangle

upward if in y, left to right if in x.

$$\text{Volume} = \pi \int_a^b (\text{radius})^2 \, dx$$

$$\text{Volume} = \pi \int_a^b (\text{radius})^2 \, dy$$

Washer Method :

outer radius is distance from the rotation axis to the far end of the rectangle

inner radius is distance from the rotation axis to the close end of the rectangle

bounds found from moving the rectangle upward if in y, left to right if in x.

$$\text{Volume} = \pi \int_a^b (\text{outer radius})^2 - (\text{inner radius})^2 \, dy$$

$$\text{Volume} = \pi \int_a^b (\text{outer radius})^2 - (\text{inner radius})^2 \, dy$$

Shell Method :

connect the rectangle to the rotation axis, this distance is the radius

the height is how long the rectangle is

bounds found from moving upward if in y, left to right if in x.

$$\text{Volume} = 2\pi \int_a^b (\text{radius})(\text{height}) \, dx \quad \text{or} \quad \text{Volume} = 2\pi \int_a^b (\text{radius})(\text{height}) \, dy$$

> **Important :**
> The distance off the x – axis $= y$
> The distance off the y – axis $= x$
> Height of the rectangle found by subtracting
> $(\text{upper}) - (\text{lower})$ if in x or $(\text{right}) - (\text{left})$ if in y

Arclength

(1) Take y'

  (1.5) Simplify

(2) Find $(y')^2$

  (2.5) Simplify

  > **Note :**
  > If $(y')^2$ is a fraction say $\frac{a}{b}$,
  > then when you add 1 add it as $\frac{b}{b}$

(3) Find $1 + (y')^2$

  (3.5) Simplify try to represent
  it as a perfect square

(4) Find $\sqrt{1 + (y')^2}$

(5) Integrate: $\text{Arclength} = \int_a^b \sqrt{1 + (y')^2} \, dx$

Surface Area

(1) Recognize the rotation axis

  (i) Rotation axis $= x \Rightarrow$ Surface Area $= 2\pi \int y \, ds$

  (ii) Rotation axis $= y \Rightarrow$ Surface Area $= 2\pi \int x \, ds$

(2) Pick your integration variable

  (i) Given $y = f(x) \ a \le x \le b \quad \Rightarrow ds = \sqrt{1 + (y'(x))^2} \, dx$
  $\Rightarrow$ do the integral in x — do the arclength steps

  (ii) Given $x = g(y) \ c \le y \le d \quad \Rightarrow ds = \sqrt{1 + (x'(y))^2} \, dy$
  $\Rightarrow$ do the integral in y — do the arclength steps

If the Rotation axis $= x \Rightarrow$ Surface Area $= 2\pi \int y \, ds$

(i) and you are integrating in x, then y needs to be replaced by $f(x)$

(ii) and you are integrating in y, then leave the y inside the integral

If the Rotation axis $= y \Rightarrow$ Surface Area $= 2\pi \int x \, ds$

(i) and you are integrating in y, then x needs to be replaced by $g(y)$

(ii) and you are integrating in x, then leave the x inside the integral

Find the length of the curve.

$$y = x^2 - \frac{\ln x}{8} \qquad \text{for } 1 \le x \le 2.$$

$$\boxed{\text{Arc Length} = \int_a^b \sqrt{1 + \left[f'(x) \right]^2}\, dx}$$

Find the derivative.

$$\Rightarrow y' = 2x - \frac{1}{8x}$$

Square the derivative.

$$\left(y' \right)^2 = \left(2x - \frac{1}{8x} \right)\left(2x - \frac{1}{8x} \right) = 4x^2 - \frac{1}{4} - \frac{1}{4} + \frac{1}{64x^2}$$

$$\Rightarrow \left(y' \right)^2 = \boxed{4x^2 - \frac{1}{2} + \frac{1}{64x^2}}$$

Add 1 and attempt to represent the expression as a perfect square.

$$1 + \left(y' \right)^2 = 4x^2 - \frac{1}{2} + 1 + \frac{1}{64x^2} = \boxed{4x^2 + \frac{1}{2} + \frac{1}{64x^2}} = \left(2x + \frac{1}{8x} \right)\left(2x + \frac{1}{8x} \right)$$

Take the square root and integrate.

$$\sqrt{1 + \left(y' \right)^2} = \sqrt{\left(2x + \frac{1}{8x} \right)^2} = 2x + \frac{1}{8x}$$

$$\int_1^2 \sqrt{1 + \left(y' \right)^2}\, dx = \int_1^2 \left(2x + \frac{1}{8x} \right) dx = \left[x^2 + \frac{\ln x}{8} \right]_1^2 = \left(4 + \frac{\ln 2}{8} \right) - (1 + 0) = \boxed{3 + \frac{1}{8}\ln 2}$$

Find the area of surface obtained by rotating the graph of a function about a coordinate axis.

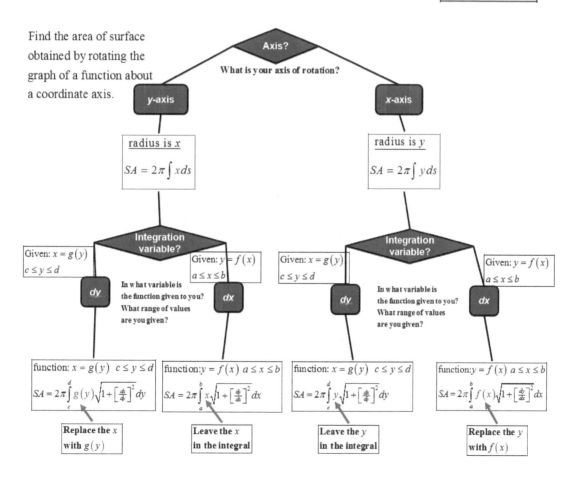

Center of Mass

General Formulas

upper   lower

Thin plate : region between $y = f(x)$ and $y = g(x)$ with $f(x) \ge g(x)$

Constant density function $\rho(x) = \rho$

Set $g(x) = 0$ in the previous formulas.

Moment about the $y - axis$

$$M_y = \rho \int_a^b x \cdot (f(x) - g(x)) dx$$

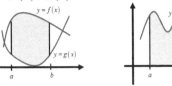

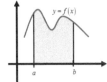

Moment about the $x - axis$

$$M_x = \frac{\rho}{2} \int_a^b \left(\left[f(x) \right]^2 - \left[g(x) \right]^2 \right) dx$$

Center of Mass

$(\overline{x}, \overline{y})$

Mass

$$M = \rho \cdot \int_a^b (f(x) - g(x)) dx$$

$$\overline{x} = \frac{M_y}{M} \qquad \overline{y} = \frac{M_x}{M}$$

If constant density function $\rho(x) = \rho$, you can ignore it since it cancels out

If density is not a constant function, you **CANNOT** ignore it, since you cannot factor it out of the integral.

Center of mass for familiar polygons

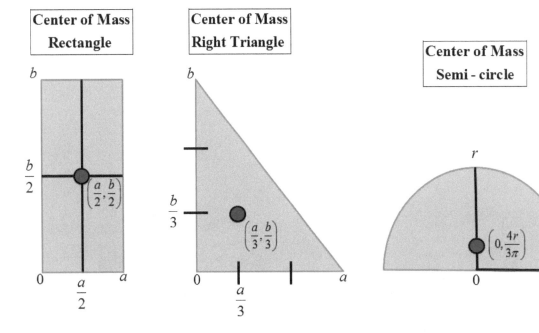

Center of Mass Rectangle

b

$\dfrac{b}{2}$

$\left(\dfrac{a}{2}, \dfrac{b}{2} \right)$

$0 \quad \dfrac{a}{2} \quad a$

Center of Mass Right Triangle

b

$\dfrac{b}{3}$

$\left(\dfrac{a}{3}, \dfrac{b}{3} \right)$

$0 \quad \dfrac{a}{3} \quad a$

Center of Mass Semi - circle

r

$\left(0, \dfrac{4r}{3\pi} \right)$

$0 \quad r$

1.1 Find the volume of the solid that lies between planes perpendicular to the x-axis at $x = 0$ and $x = 1$. The cross sections of the solid perpendicular to the x-axis between these planes are semicircles whose diameters run from the curve $y = x^3$ to the curve $y = x$.

(A) $\dfrac{\pi}{16}$ (C) $\dfrac{\pi}{60}$ (E) $\dfrac{\pi}{40}$ (G) $\dfrac{\pi}{105}$

(B) $\dfrac{\pi}{80}$ (D) $\dfrac{\pi}{24}$ (F) $\dfrac{\pi}{240}$ (H) $\dfrac{\pi}{120}$

1.1 Find the volume of the solid that lies between planes perpendicular to the x – axis at $x = 0$ and $x = 1$. The cross sections of the solid perpendicular to the x – axis between these planes are semicircles whose diameters run from the curve $y = x^3$ to the curve $y = x$.

(A) $\dfrac{\pi}{16}$ (C) $\dfrac{\pi}{60}$ (E) $\dfrac{\pi}{40}$ (G) $\dfrac{\pi}{105}$

(B) $\dfrac{\pi}{80}$ (D) $\dfrac{\pi}{24}$ (F) $\dfrac{\pi}{240}$ (H) $\dfrac{\pi}{120}$

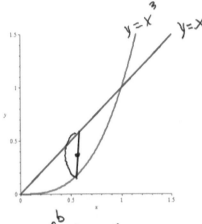

The diameter d is the distance between the curves

$$d = x - x^3$$

The radius r is half of the distance between the curves

$$r = \frac{x - x^3}{2}$$

The shape is a semicircle so the area A is

$$A(r) = \frac{1}{2}\pi r^2$$

$$A(x) = \frac{\pi}{2}\left(\frac{x - x^3}{2}\right)^2$$

$$A(x) = \frac{\pi}{2}\cdot\frac{1}{4}\left(x - x^3\right)^2$$

$$A(x) = \frac{\pi}{8}\left(x - x^3\right)\left(x - x^3\right)$$

$$\boxed{A(x) = \frac{\pi}{8}\left(x^2 - 2x^4 + x^6\right)}$$

$$V = \int_a^b A(x)\,dx$$
cross-sectional area formula

$$V = \int_0^1 \frac{\pi}{8}\left(x^2 - 2x^4 + x^6\right)dx$$

$$V = \frac{\pi}{8}\left[\frac{x^3}{3} - \frac{2x^5}{5} + \frac{x^7}{7}\right]_0^1$$

$$V = \frac{\pi}{8}\left[\frac{1}{3} - \frac{2}{5} + \frac{1}{7}\right] = \frac{\pi}{8}\left[\frac{1\cdot 35}{3\cdot 35} - \frac{2\cdot 21}{5\cdot 21} + \frac{1\cdot 15}{7\cdot 15}\right]$$

$$V = \frac{\pi}{8}\left[\frac{35 - 42 + 15}{105}\right] = \frac{\pi}{8}\left[\frac{50 - 42}{105}\right] = \frac{\pi}{8}\left[\frac{8}{105}\right]$$

$$\boxed{V = \frac{\pi}{105}\text{ units}^3}$$

1.2 Find the volume of the solid generated by revolving the region in the first quadrant bounded by $y = x^2 + 1, x = 0,$ and $y = 2$ about the line $y = 2$

(A) $\dfrac{4\pi}{5}$

(C) $\dfrac{8\pi}{15}$

(E) $\dfrac{12\pi}{5}$

(G) $\dfrac{4\pi}{15}$

(B) $\dfrac{7\pi}{10}$

(D) $\dfrac{2\pi}{3}$

(F) $\dfrac{2\pi}{15}$

(H) $\dfrac{2\pi}{5}$

1.2 Find the volume of the solid generated by revolving the region in the first quadrant bounded by $y = x^2 + 1, x = 0,$ and $y = 2$ about the line $y = 2$

(A) $\dfrac{4\pi}{5}$ (C) $\dfrac{8\pi}{15}$ (E) $\dfrac{12\pi}{5}$ (G) $\dfrac{4\pi}{15}$

(B) $\dfrac{7\pi}{10}$ (D) $\dfrac{2\pi}{3}$ (F) $\dfrac{2\pi}{15}$ (H) $\dfrac{2\pi}{5}$

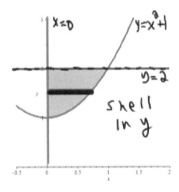

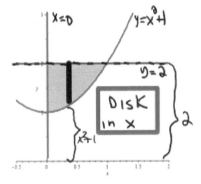

Find where the curves intersect

$$x^2 + 1 = 2$$
$$x^2 = 1$$
$$x = \pm 1$$

The radius r is

$$r = 2 - (x^2 + 1)$$
$$r = 2 - x^2 - 1$$
$$r = 1 - x^2$$

$$V_{disk} = \pi \int_a^b [F(x)]^2 \, dx$$

$$V = \pi \int_0^1 (1 - x^2)^2 \, dx$$

$$V = \pi \int_0^1 (1 - 2x^2 + x^4) \, dx$$

$$V = \pi \left[x - \frac{2x^3}{3} + \frac{x^5}{5} \right]_0^1$$

$$V = \pi \left[1 - \frac{2}{3} + \frac{1}{5} \right]$$

$$V = \pi \left[\frac{1}{3} + \frac{1}{5} \right]$$

$$V = \pi \left[\frac{5 + 3}{15} \right]$$

$$\boxed{V = \frac{8\pi}{15} \text{ units}^3}$$

1.3 Find the volume of the solid generated when the region bounded by $y = x^2 + 1$, $y = -x + 1$ and $x = 1$ is revolved about the $y-$axis.

(A) 1

(C) $\dfrac{\pi}{2}$

(E) $\dfrac{7\pi}{6}$

(G) $\dfrac{\pi}{8}$

(B) π

(D) $\dfrac{\pi}{6}$

(F) $\dfrac{\pi}{4}$

(H) $\dfrac{5\pi}{6}$

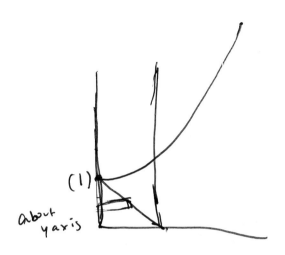

(1)

about
y axis

$$2 \int_{0}^{1} 1^2 - (-x+1) + \left(x^2\right)^2 -$$

1.3 Find the volume of the solid generated when the region bounded by $y = x^2 + 1$, $y = -x + 1$ and $x = 1$ is revolved about the y – axis.

(A) 1

(C) $\dfrac{\pi}{2}$

(E) $\dfrac{7\pi}{6}$

(G) $\dfrac{\pi}{8}$

(B) π

(D) $\dfrac{\pi}{6}$

(F) $\dfrac{\pi}{4}$

(H) $\dfrac{5\pi}{6}$

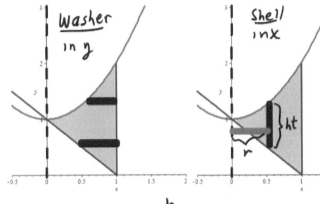

The radius r is the distance off the x-axis

$r = x$

The height ht is the difference between the parabola and the line

$ht = (x^2 + 1) - (-x + 1)$

$ht = x^2 + \cancel{1} + x - \cancel{1}$

$ht = x^2 + x$

$$V_{Shell} = 2\pi \int_a^b (radius)(height)\, dx$$

$$V = 2\pi \int_0^1 x(x^2 + x)\, dx = 2\pi \int_0^1 (x^3 + x^2)\, dx$$

$$V = 2\pi \left[\frac{x^4}{4} + \frac{x^3}{3} \right]_0^1 = 2\pi \left(\frac{1}{4} + \frac{1}{3} \right)$$

$$V = 2\pi \left[\frac{3 + 4}{12} \right] = 2\pi \left[\frac{7}{12} \right]$$

$$\boxed{V = \frac{7\pi}{6} \text{ units}^3}$$

1.4 Find the volume of the solid generated by revolving the region bounded by $y = 2 + \sqrt{x-1}, x = 2, x = 5,$ and $y = 2,$ about the $x-$axis.

(A) $\dfrac{17\pi}{6}$

(B) 24π

(C) $\dfrac{11\pi}{2}$

(D) $\dfrac{145\pi}{6}$

(E) $\dfrac{15\pi}{2}$

(F) $\dfrac{151\pi}{6}$

(G) 40π

(H) $\dfrac{157\pi}{6}$

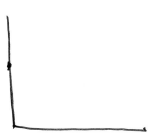

- Volume / SA
- Trig Sub
- converge / Divergence
- Probability

1.4 Find the volume of the solid generated by revolving the region bounded by $y = 2 + \sqrt{x-1}, x = 2, y = 5,$ and $y = 2,$ about the $x-$axis.

(A) $\dfrac{17\pi}{6}$ (C) $\dfrac{11\pi}{2}$ (E) $\dfrac{15\pi}{2}$ (G) 40π

(B) 24π (D) $\dfrac{145\pi}{6}$ (F) $\dfrac{151\pi}{6}$ (H) $\dfrac{157\pi}{6}$

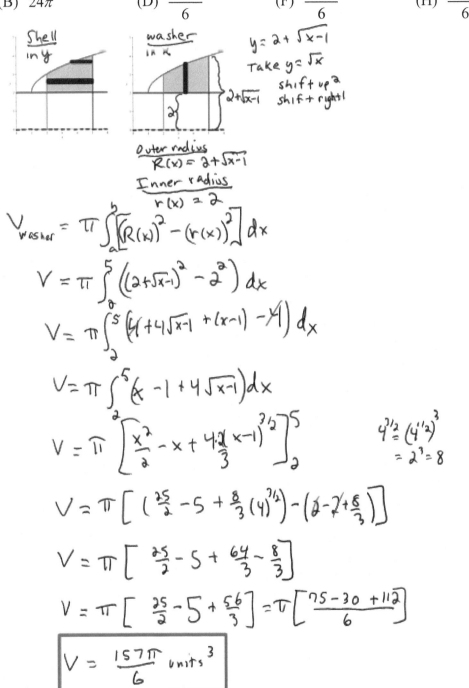

Shell in y

washer in x

$y = 2 + \sqrt{x-1}$

Take $y = \sqrt{x}$
shift up 2
$2 + \sqrt{x-1}$ shift right 1

Outer radius
$\underline{R(x) = 2 + \sqrt{x-1}}$
Inner radius
$\underline{r(x) = 2}$

$$V_{washer} = \pi \int_a^b \left[(R(x))^2 - (r(x))^2 \right] dx$$

$$V = \pi \int_2^5 \left((2+\sqrt{x-1})^2 - 2^2 \right) dx$$

$$V = \pi \int_2^5 \left(4 + 4\sqrt{x-1} + (x-1) - 4 \right) dx$$

$$V = \pi \int_2^5 \left(x - 1 + 4\sqrt{x-1} \right) dx$$

$$V = \pi \left[\frac{x^2}{2} - x + \frac{4\cdot 2}{3}(x-1)^{3/2} \right]_2^5$$

$4^{3/2} = (4^{1/2})^3$
$= 2^3 = 8$

$$V = \pi \left[\left(\frac{25}{2} - 5 + \frac{8}{3}(4)^{3/2} \right) - \left(2 - 2 + \frac{8}{3} \right) \right]$$

$$V = \pi \left[\frac{25}{2} - 5 + \frac{64}{3} - \frac{8}{3} \right]$$

$$V = \pi \left[\frac{25}{2} - 5 + \frac{56}{3} \right] = \pi \left[\frac{75 - 30 + 112}{6} \right]$$

$$\boxed{V = \frac{157\pi}{6} \text{ units}^3}$$

1.5 Find the volume of the solid generated by revolving the region bounded by $y = 4 - x^2$, $y = 4$, and $x = 2$, about the x-axis.

(A) $\dfrac{74\pi}{5}$

(B) $\dfrac{115\pi}{9}$

(C) $\dfrac{97\pi}{12}$

(D) $\dfrac{85\pi}{6}$

(E) $\dfrac{124\pi}{15}$

(F) $\dfrac{109\pi}{9}$

(G) $\dfrac{224\pi}{15}$

(H) $\dfrac{152\pi}{5}$

1.5 Find the volume of the solid generated by revolving the region bounded by $y = 4 - x^2$, $y = 4$, and $x = 2$, about the $x-$ axis.

(A) $\dfrac{74\pi}{5}$

(C) $\dfrac{97\pi}{12}$

(E) $\dfrac{124\pi}{15}$

(G) $\dfrac{224\pi}{15}$

(B) $\dfrac{115\pi}{9}$

(D) $\dfrac{85\pi}{6}$

(F) $\dfrac{109\pi}{9}$

(H) $\dfrac{152\pi}{5}$

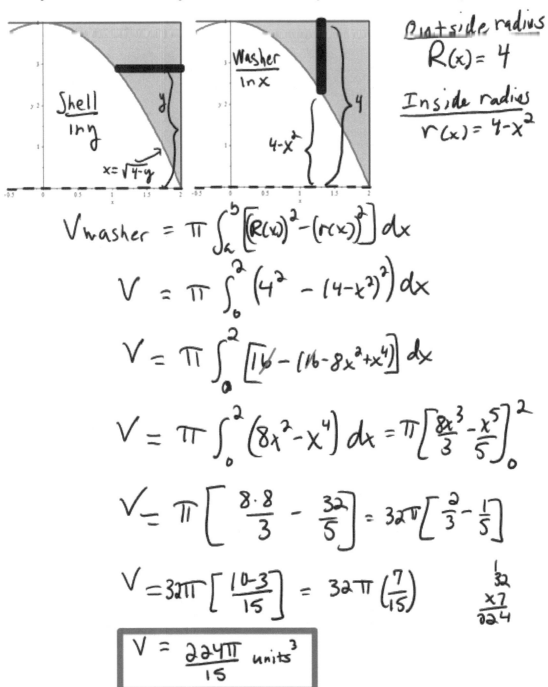

Outside radius
$$R(x) = 4$$

Inside radius
$$r(x) = 4 - x^2$$

$$V_{washer} = \pi \int_a^b \left[\left(R(x)\right)^2 - \left(r(x)\right)^2 \right] dx$$

$$V = \pi \int_0^2 \left(4^2 - (4-x^2)^2 \right) dx$$

$$V = \pi \int_0^2 \left[16 - (16 - 8x^2 + x^4) \right] dx$$

$$V = \pi \int_0^2 \left(8x^2 - x^4 \right) dx = \pi \left[\frac{8x^3}{3} - \frac{x^5}{5} \right]_0^2$$

$$V = \pi \left[\frac{8 \cdot 8}{3} - \frac{32}{5} \right] = 32\pi \left[\frac{2}{3} - \frac{1}{5} \right]$$

$$V = 32\pi \left[\frac{10-3}{15} \right] = 32\pi \left(\frac{7}{15} \right)$$

$$\boxed{V = \frac{224\pi}{15} \text{ units}^3}$$

$$\begin{array}{r} 32 \\ \times 7 \\ \hline 224 \end{array}$$

1.6 Find the volume of the solid generated by revolving the region bounded by $y = \ln x$, $y = 0$, and $x = e$ about the $x-$axis.

(A) $\frac{\pi}{2}(e+1)$

(C) $\pi(4-e)$

(E) $\frac{\pi}{4}e+2\pi$

(G) π

(B) 2π

(D) $\frac{\pi}{2}e$

(F) $\pi(e-2)$

(H) πe

1.6 Find the volume of the solid generated by revolving the region bounded by $y = \ln x$, $y = 0$, and $x = e$ about the $x-$axis.

(A) $\dfrac{\pi}{2}(e+1)$ (C) $\pi(4-e)$ (E) $\dfrac{\pi}{4}e+2\pi$ (G) π

(B) 2π (D) $\dfrac{\pi}{2}e$ (F) $\pi(e-2)$ (H) πe

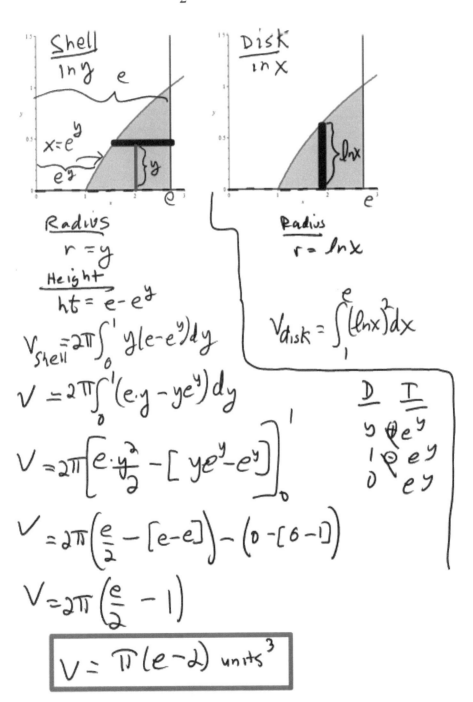

Shell
in y

Disk
in x

$x = e^{y}$

e^{y}

Radius
$r = y$

Height
$ht = e - e^{y}$

$V_{shell} = 2\pi \int_{0}^{1} y(e - e^{y})dy$

$V = 2\pi \int_{0}^{1} (e \cdot y - y e^{y})dy$

$V = 2\pi \left[e \cdot \dfrac{y^{2}}{2} - \left[y e^{y} - e^{y} \right] \right]_{0}^{1}$

$V = 2\pi \left(\dfrac{e}{2} - [e - e] \right) - (0 - [0 - 1])$

$V = 2\pi \left(\dfrac{e}{2} - 1 \right)$

Radius
$r = \ln x$

$V_{disk} = \int_{1}^{e} (\ln x)^{2} dx$

$\dfrac{D}{y} \quad \dfrac{I}{\theta e^{y}}$
$1 \quad e^{y}$
$0 \quad e^{y}$

$\boxed{V = \pi(e - 2) \text{ units}^{3}}$

1.7 **Setup but DON'T solve.** Find the volume of the solid generated by revolving the region bounded by $y = \sqrt{x+3}$, $y = 0$, and $x + y = 3$, about the line $y = -2$. **Setup but DON'T solve.**

1.7 Setup but DON'T solve. Find the volume of the solid generated by revolving the region bounded by $y = \sqrt{x+3}$, $y = 0$, and $x + y = 3$, about the line $y = -2$. **Setup but DON'T solve.**

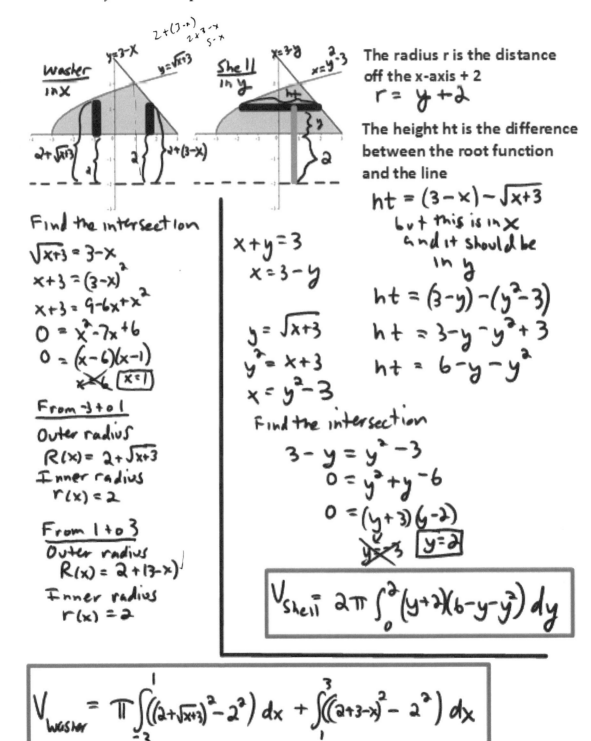

Washer in x

$y = 3-x$ $y = \sqrt{x+3}$

$2 + (3-x)$
$\quad \frac{2 \times 3 - x}{5 - x}$

Shell in y

$x = 3-y$ $x = y^2 - 3$

The radius r is the distance off the x-axis + 2

$$r = y + 2$$

The height ht is the difference between the root function and the line

$$ht = (3-x) - \sqrt{x+3}$$
but this is in x and it should be in y

$$ht = (3-y) - (y^2 - 3)$$
$$ht = 3 - y - y^2 + 3$$
$$ht = 6 - y - y^2$$

Find the intersection

$\sqrt{x+3} = 3-x$
$x+3 = (3-x)^2$
$x+3 = 9 - 6x + x^2$
$0 = x^2 - 7x + 6$
$0 = (x-6)(x-1)$
$\cancel{x=6} \quad \boxed{x=1}$

From -3 to 1
Outer radius
$R(x) = 2 + \sqrt{x+3}$
Inner radius
$r(x) = 2$

From 1 to 3
Outer radius
$R(x) = 2 + (3-x)$
Inner radius
$r(x) = 2$

$x + y = 3$
$x = 3 - y$

$y = \sqrt{x+3}$
$y^2 = x + 3$
$x = y^2 - 3$

Find the intersection

$3 - y = y^2 - 3$
$0 = y^2 + y - 6$
$0 = (y+3)(y-2)$
$\cancel{y = -3} \quad \boxed{y = 2}$

$$V_{Shell} = 2\pi \int_0^2 (y+2)(6-y-y^2)\, dy$$

$$V_{washer} = \pi \int_{-3}^{1} \left((2+\sqrt{x+3})^2 - 2^2\right) dx + \int_1^3 \left((2+3-x)^2 - 2^2\right) dx$$

1.8 The base of a solid is in the first quadrant between the curve $y = x^2$ and the curve $y = \sqrt{x}$ for $0 \le x \le 1$. The cross sections of the solid perpendicular to the x – axis are isosceles right triangles whose leg runs between the curves. Find the volume of the solid.

(A) $\dfrac{3}{70}$ (C) $\dfrac{\sqrt{2}}{25}$ (E) $\dfrac{9}{35}$ (G) $\dfrac{37}{70}$

(B) $\dfrac{3}{35}$ (D) $\dfrac{9}{140}$ (F) $\dfrac{9}{70}$ (H) $\dfrac{9\sqrt{3}}{280}$

1.8 The base of a solid is in the first quadrant between the curve $y = x^2$ and the curve $y = \sqrt{x}$
for $0 \le x \le 1$. The cross sections of the solid perpendicular to the $x-$axis are isosceles right
triangles whose leg runs between the curves. Find the volume of the solid.

(A) $\dfrac{3}{70}$ (C) $\dfrac{\sqrt{2}}{25}$ (E) $\dfrac{9}{35}$ (G) $\dfrac{37}{70}$

(B) $\dfrac{3}{35}$ (D) $\dfrac{9}{140}$ (F) $\dfrac{9}{70}$ (H) $\dfrac{9\sqrt{3}}{280}$

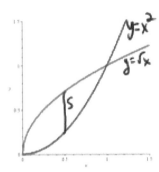

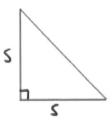

$V = \int_{r}^{n} a(x)$

The leg length s is the
difference between
the root function
and the parabola

$S = \sqrt{x} - x^2$

$A(s) = \frac{1}{2} \cdot s \cdot s$

$A(s) = \frac{1}{2} s^2$

$A(x) = \frac{1}{2}\left(\sqrt{x} - x^2\right)^2$

$A(x) = \frac{1}{2}\left(x - 2\sqrt{x}x^2 + x^4\right)$

$A(x) = \frac{1}{2}\left(x - 2x^{5/2} + x^4\right)$

$V = \int_{a}^{b} A(x)\, dx = \int_{0}^{1} \frac{1}{2}\left(x - 2x^{5/2} + x^4\right) dx$

$V = \frac{1}{2}\left[\dfrac{x^2}{2} - 2 \cdot \frac{2}{7} x^{7/2} + \dfrac{x^5}{5}\right]_{0}^{1}$

$V = \frac{1}{2}\left[\left(\frac{1}{2} - \frac{4}{7} + \frac{1}{5}\right) - 0\right]$

$V = \frac{1}{2}\left[\dfrac{35 - 40 + 14}{70}\right]$

$V = \dfrac{9}{140}$ units^3

28

1.9 Find the volume of the solid generated by revolving the region bounded by $y = x$, $y = -x$, and $x = 2$, about the line $x = -3$.

(A) $\dfrac{74\pi}{4}$

(B) $\dfrac{115\pi}{3}$

(C) $\dfrac{97\pi}{12}$

(D) $\dfrac{85\pi}{6}$

(E) $\dfrac{26\pi}{3}$

(F) $\dfrac{104\pi}{3}$

(G) $\dfrac{208\pi}{3}$

(H) $\dfrac{52\pi}{9}$

1.9 Find the volume of the solid generated by revolving the region bounded by $y = x$, $y = -x$, and $x = 2$, about the line $x = -3$.

(A) $\dfrac{74\pi}{4}$ (C) $\dfrac{97\pi}{12}$ (E) $\dfrac{26\pi}{3}$ (G) $\dfrac{208\pi}{3}$

(B) $\dfrac{115\pi}{3}$ (D) $\dfrac{85\pi}{6}$ (F) $\dfrac{104\pi}{3}$ (H) $\dfrac{52\pi}{9}$

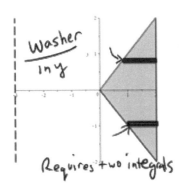

Washer
in y

Requires two integrals

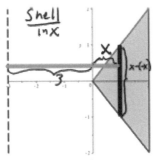

Shell
in x

The radius r is the distance off the y-axis + 3

$$r = x + 3$$

The height ht is the difference between the line y=x and y=-x

$$ht = x - (-x)$$
$$ht = x + x$$
$$ht = 2x$$

$$V_{Shell} = 2\pi \int_a^b (radius)(height)\, dx$$

$$= 2\pi \int_0^2 (x+3)(2x)\, dx$$

$$= 4\pi \int_0^2 (x^2 + 3x)\, dx$$

$$= 4\pi \left[\frac{x^3}{3} + \frac{3x^2}{2} \right]_0^2$$

$$V = 4\pi \left[\left(\frac{8}{3} + 3\cdot 2 \right) - 0 \right]$$

$$V = 4\pi \left[\frac{8}{3} + 6 \right] = 4\pi \left(\frac{8+18}{3} \right)$$

$$\boxed{V = \frac{104\pi}{3}\ units^3}$$

2
26
×4
104

1.10 Find the volume of the solid generated by revolving the region bounded by $y = \sqrt{x}$, $x = 0$, and $y = 1$ about the line $y = -1$.

(A) $\dfrac{7\pi}{6}$

(B) $\dfrac{5\pi}{3}$

(C) $\dfrac{5\pi}{12}$

(D) $\dfrac{9\pi}{5}$

(E) $\dfrac{5\pi}{4}$

(F) $\dfrac{7\pi}{3}$

(G) $\dfrac{5\pi}{6}$

(H) $\dfrac{7\pi}{12}$

1.10 Find the volume of the solid generated by revolving the region bounded by $y = \sqrt{x}$, $x = 0$, and $y = 1$ about the line $y = -1$.

(A) $\dfrac{7\pi}{6}$ (C) $\dfrac{5\pi}{12}$ (E) $\dfrac{5\pi}{4}$ (G) $\dfrac{5\pi}{6}$

(B) $\dfrac{5\pi}{3}$ (D) $\dfrac{9\pi}{5}$ (F) $\dfrac{7\pi}{3}$ (H) $\dfrac{7\pi}{12}$

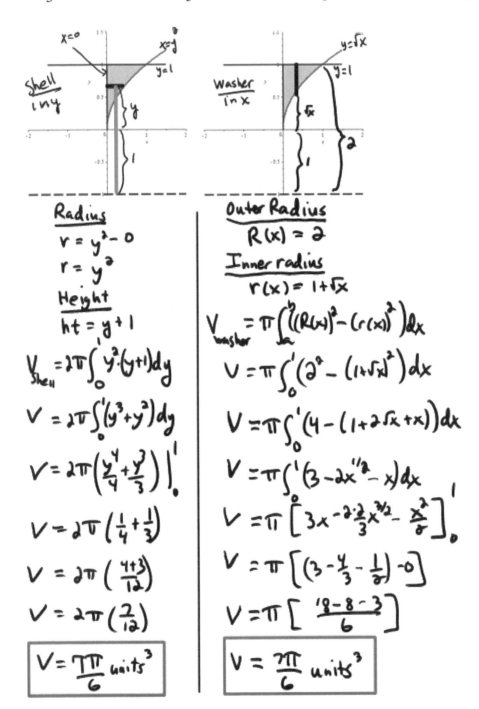

Radius
$$r = y^2 - 0$$
$$r = y^2$$

Height
$$ht = y + 1$$

$$V_{Shell} = 2\pi \int_0^1 y^2 (y+1) \, dy$$

$$V = 2\pi \int_0^1 (y^3 + y^2) \, dy$$

$$V = 2\pi \left(\frac{y^4}{4} + \frac{y^3}{3} \right) \Big|_0^1$$

$$V = 2\pi \left(\frac{1}{4} + \frac{1}{3} \right)$$

$$V = 2\pi \left(\frac{4+3}{12} \right)$$

$$V = 2\pi \left(\frac{7}{12} \right)$$

$$\boxed{V = \frac{7\pi}{6} \text{ units}^3}$$

Outer Radius
$$R(x) = 2$$

Inner radius
$$r(x) = 1 + \sqrt{x}$$

$$V_{washer} = \pi \int_a^b \left((R(x))^2 - (r(x))^2 \right) \, dx$$

$$V = \pi \int_0^1 \left(2^2 - (1+\sqrt{x})^2 \right) \, dx$$

$$V = \pi \int_0^1 \left(4 - (1 + 2\sqrt{x} + x) \right) \, dx$$

$$V = \pi \int_0^1 (3 - 2x^{1/2} - x) \, dx$$

$$V = \pi \left[3x - 2 \cdot \frac{2}{3} x^{3/2} - \frac{x^2}{2} \right]_0^1$$

$$V = \pi \left[\left(3 - \frac{4}{3} - \frac{1}{2} \right) - 0 \right]$$

$$V = \pi \left[\frac{18 - 8 - 3}{6} \right]$$

$$\boxed{V = \frac{7\pi}{6} \text{ units}^3}$$

1.11 Find the volume of the solid generated by revolving the region bounded by $y = \sqrt{x}\left(x^2 + 16\right)^{1/4}$, $x = 3$ and $y = 0$, about the $x-$axis. See the graph below.

(A) $\dfrac{17\pi}{3}$

(B) $\dfrac{19\pi}{3}$

(C) $\dfrac{23\pi}{3}$

(D) $\dfrac{53\pi}{3}$

(E) $\dfrac{61\pi}{3}$

(F) $\dfrac{47\pi}{3}$

(G) $\dfrac{67\pi}{3}$

(H) $\dfrac{81\pi}{2}$

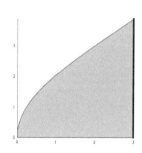

1.11 Find the volume of the solid generated by revolving the region bounded by $y = \sqrt{x}\left(x^2 + 16\right)^{1/4}$, $x = 3$ and $y = 0$, about the $x-$axis. See the graph below.

(A) $\dfrac{17\pi}{3}$ (C) $\dfrac{23\pi}{3}$ (E) $\dfrac{61\pi}{3}$ (G) $\dfrac{67\pi}{3}$

(B) $\dfrac{19\pi}{3}$ (D) $\dfrac{53\pi}{3}$ (F) $\dfrac{47\pi}{3}$ (H) $\dfrac{81\pi}{2}$

$$V_{disk} = \pi \int_a^b \left(r(x)\right)^2 dx$$

$$V = \pi \int_0^3 \left(\sqrt{x}(x^2+16)^{1/4}\right)^2 dx$$

$$V = \pi \int_0^3 x \cdot \sqrt{x^2+16}\; dx$$

$$V = \frac{\pi}{3}\left(x^2+16\right)^{3/2}\Big|_0^3$$

$$V = \frac{\pi}{3}\left[(9+16)^{3/2} - 16^{3/2}\right]$$

$$V = \frac{\pi}{3}(125 - 64)$$

$$\boxed{V = \frac{61\pi}{3}\;\text{units}^3}$$

DISK in X

$radius = \sqrt{x}\,(x^2+16)^{1/4}$

$u = x^2+16$
$du = 2x\,dx$
$\frac{1}{2}du = x\,dx$

$\frac{1}{2}\int u^{1/2}\,du$

$\frac{1}{2}\frac{2}{3}u^{3/2}$

$25^{3/2} = \left(\sqrt{25}\right)^3$
$= 5^3$
$= 125$

$16^{3/2} = \left(\sqrt{16}\right)^3$
$= 4^3$
$= 64$

1.12 Find the $x-$value of the point P on $y = \dfrac{2}{3}x^{3/2}$ to the right of the

$y-$axis so that the length of the curve from $(0,0)$ to P is $\dfrac{52}{3}$.

(A) 8   (C) 9   (E) 6   (G) 21

(B) 2   (D) 4   (F) 15   (H) 27

1.12 Find the x – value of the point P on $y = \dfrac{2}{3}x^{3/2}$ to the right of the

y – axis so that the length of the curve from $(0,0)$ to P is $\dfrac{52}{3}$.

(A) 8   (C) 9   (E) 6   (G) 21

(B) 2   (D) 4   (F) 15   (H) 27

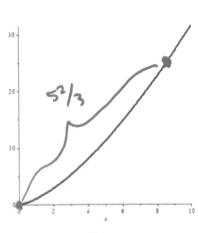

$$y = \tfrac{2}{3}x^{3/2}$$

$$y' = \tfrac{2}{3} \cdot \tfrac{3}{2} x^{1/2} = \sqrt{x}$$

$$(y')^2 = x$$

$$1 + (y')^2 = 1 + x$$

$$\text{Arc length} = \int_{a}^{b} \sqrt{1 + (y')^2}\, dx$$

$$\frac{52}{3} = \int_{0}^{b} \sqrt{1+x}\, dx$$

$$\frac{52}{3} = \left[\tfrac{2}{3}(1+x)^{3/2} \, \Big|_{0}^{b} \right]$$

$$\frac{52}{3} = \tfrac{2}{3}\left[(1+b)^{3/2} - 1 \right]$$

$$\frac{52}{3} \cdot \tfrac{3}{2} \cdot \frac{1}{ } = (1+b)^{3/2} - 1$$

$$26 + 1 = (1+b)^{3/2}$$

$$27^{2/3} = 1+b$$

$$\left(27^{1/3}\right)^2 = 1+b$$

$$3^2 - 1 = b$$

$$\Rightarrow \boxed{b = 8}$$

1.13 Find the arclength of the curve $y = \left(4 - x^{2/3}\right)^{3/2}$ for $1 \le x \le 8$.

(A) 8   (C) 9   (E) 6   (G) 21

(B) 2   (D) 4   (F) 15   (H) 27

1.13 Find the arclength of the curve $y = \left(4 - x^{2/3}\right)^{3/2}$ for $1 \le x \le 8$.

(A) 8   (C) 9   (E) 6   (G) 21

(B) 2   (D) 4   (F) 15   (H) 27

$$\text{Arclength} = \int_a^b \sqrt{1 + (y'(x))^2}\, dx$$

$$AL = \int_1^8 2x^{-1/3}\, dx$$

$$AL = \left[2 \cdot \frac{x^{2/3}}{2/3}\right]_1^8$$

$$AL = 2 \cdot \frac{3}{2}\left[8^{2/3} - 1\right]$$

$$AL = 3\left[(8^{1/3})^2 - 1\right]$$

$$AL = 3\left[2^2 - 1\right]$$

$$AL = 3 \cdot 3$$

$$\boxed{AL = 9 \text{ units}}$$

$$y = (4 - x^{2/3})^{3/2}$$

$$y' = \frac{3}{2}(4 - x^{2/3})^{1/2} \cdot -\frac{2}{3}x^{-1/3}$$

$$y' = \frac{-\sqrt{4 - x^{2/3}}}{x^{1/3}}$$

$$(y')^2 = \frac{4 - x^{2/3}}{x^{2/3}}$$

$$(y')^2 = \frac{4}{x^{2/3}} - 1$$

$$1 + (y')^2 = 1 + \left(\frac{4}{x^{2/3}} - 1\right)$$

$$1 + (y')^2 = \frac{4}{x^{2/3}}$$

$$\sqrt{1 + (y')^2} = \sqrt{\frac{4}{x^{2/3}}} = \frac{2}{x^{1/3}}$$

1.14 Let $y = 2 \cdot \ln\left(\sec\left(\dfrac{x}{2}\right)\right)$. Find the arclength for $0 \le x \le \dfrac{\pi}{3}$.

(A) $\dfrac{1}{3}$

(B) $\dfrac{1}{9}$

(C) $\ln\left(\dfrac{\sqrt{3}}{2}\right)$

(D) $\dfrac{1}{6}$

(E) $\ln 2$

(F) $\dfrac{1}{2}$

(G) $\ln 3$

(H) None of these

1.14 Let $y = 2 \cdot \ln\left(\sec\left(\dfrac{x}{2} \right) \right)$. Find the arclength for $0 \le x \le \dfrac{\pi}{3}$.

(A) $\dfrac{1}{3}$ (C) $\ln\left(\dfrac{\sqrt{3}}{2} \right)$ (E) $\ln 2$ (G) $\ln 3$

(B) $\dfrac{1}{9}$ (D) $\dfrac{1}{6}$ (F) $\dfrac{1}{2}$ (H) None of these

$$y = 2 \ln\left(\sec\left(\tfrac{x}{2} \right) \right)$$

$$y' = 2 \cdot \frac{1}{\sec\left(\frac{x}{2}\right)} \cdot \sec\left(\tfrac{x}{2}\right) \cdot \tan\left(\tfrac{x}{2}\right) \cdot \frac{1}{2} = \tan\left(\tfrac{x}{2}\right)$$

$$(y')^2 = \tan^2\left(\tfrac{x}{2}\right)$$

$$1 + (y')^2 = 1 + \tan^2\left(\tfrac{x}{2}\right) = \sec^2\left(\tfrac{x}{2}\right)$$

$$\sqrt{1 + (y')^2} = \sec\left(\tfrac{x}{2}\right)$$

$$\int_0^{\frac{\pi}{3}} \sqrt{1 + (y')^2}\, dx = \int_0^{\frac{\pi}{3}} \sec\left(\tfrac{x}{2}\right) dx = 2\left[\ln\left| \sec\left(\tfrac{x}{2}\right) + \tan\left(\tfrac{x}{2}\right) \right| \right]_0^{\pi/3}$$

$$= 2 \cdot \left[\ln\left| \sec\tfrac{\pi}{6} + \tan\tfrac{\pi}{6} \right| - \ln\left| \sec 0 + \tan 0 \right| \right]$$

$$= 2\left[\ln\left| \tfrac{2}{\sqrt{3}} + \tfrac{1}{\sqrt{3}} \right| - 0 \right] = 2\ln\left(\tfrac{3}{\sqrt{3}} \right)$$

$$= \ln\left[\left(\tfrac{3}{\sqrt{3}} \right)^2 \right] = \ln\left(\tfrac{9}{3} \right) = \boxed{\ln 3}$$

1.15 Find the length of the curve given by $x = \dfrac{y^3}{12} + \dfrac{1}{y}$, for $1 \le y \le 2$.

(A) $\dfrac{\pi}{12}$

(B) $\dfrac{13}{12}$

(C) $\dfrac{1}{12}$

(D) $\dfrac{1}{2}$

(E) $\dfrac{\pi}{2}$

(F) 2

(G) $\dfrac{5}{12}$

(H) $\dfrac{17}{12}$

1.15 Find the length of the curve given by $x = \dfrac{y^3}{12} + \dfrac{1}{y}$, for $1 \le y \le 2$.

(A) $\dfrac{\pi}{12}$ (C) $\dfrac{1}{12}$ (E) $\dfrac{\pi}{2}$ (G) $\dfrac{5}{12}$

(B) $\dfrac{13}{12}$ (D) $\dfrac{1}{2}$ (F) 2 (H) $\dfrac{17}{12}$

$$\text{Arclength} = \int_a^b \sqrt{1 + (x'(y))^2}\, dy$$

$$x = \frac{y^3}{12} + \frac{1}{y} \quad \text{so} \quad x' = \frac{3y^2}{12} - \frac{1}{y^2} = \frac{y^2}{4} - \frac{1}{y^2}$$

$$\text{Thus } (x')^2 = \left(\frac{y^2}{4} - \frac{1}{y^2}\right)\left(\frac{y^2}{4} - \frac{1}{y^2}\right)$$

$$(x')^2 = \frac{y^4}{16} - \frac{1}{4} - \frac{1}{4} + \frac{1}{y^4}$$

$$(x')^2 = \boxed{\frac{y^4}{16} - \frac{1}{2} + \frac{1}{y^4} = \left(\frac{y^2}{4} + \frac{1}{y^2}\right)^2}$$

Now add 1

$$1 + (x')^2 = \frac{y^4}{16}\left[-\frac{1}{2} + 1\right] + \frac{1}{y^4}$$

$$1 + (x')^2 = \boxed{\frac{y^4}{16} + \frac{1}{2} + \frac{1}{y^4} = \left(\frac{y^2}{4} + \frac{1}{y^2}\right)^2}$$

If $1 + (x')^2 = \left(\frac{y^2}{4} + \frac{1}{y^2}\right)^2$, then $\sqrt{1+(x')^2} = \frac{y^2}{4} + \frac{1}{y^2}$

Finally integrate to find the arc length:

$$A.L = \int_1^2 \left(\frac{y^2}{4} + \frac{1}{y^2}\right) dy = \int_1^2 \left(\frac{y^2}{4} + y^{-2}\right) dy$$

$$A.L = \left[\frac{y^3}{12} - \frac{1}{y}\right]\Big|_1^2 = \left(\frac{8}{12} - \frac{1}{2}\right) - \left(\frac{1}{12} - 1\right)$$

$$A.L = \frac{7}{12} - \frac{1}{2} + 1 = \frac{7 - 6 + 12}{12}$$

$$\boxed{A.L = \frac{13}{12} \text{ units}}$$

1.16 Find the area of the surface generated by revolving the curve $x = \sqrt{9-y^2}$ for $-1 \le y \le 1$ about the $y-$axis.

(A) 2π (C) 4π (E) 8π (G) 12π

(B) π (D) 24π (F) 36π (H) 15π

1.16 Find the area of the surface generated by revolving the curve $x = \sqrt{9 - y^2}$ for $-1 \le y \le 1$ about the $y-$axis.

(A) 2π (C) 4π (E) 8π (G) 12π

(B) π (D) 24π (F) 36π (H) 15π

Surface Area for a curve rotated about y-axis

$$SA = \int 2\pi x \, ds$$

We are given y-bounds and the function is given as $x = g(y)$ } Do the integral in y

$$SA = 2\pi \int_a^b \underbrace{\sqrt{9-y^2}}_{x} \sqrt{1 + (x'(y))^2} \, dy = 2\pi \int_{-1}^{1} \sqrt{9-y^2} \cdot \frac{3}{\sqrt{9-y^2}} \, dy$$

$x = \sqrt{9-y^2}$

Find $x'(y)$

$$x' = \frac{1}{2\sqrt{9-y^2}} \cdot (-2y) = \frac{-y}{\sqrt{9-y^2}}$$

Now square x'

$$(x')^2 = \frac{y^2}{9-y^2}$$

Next add 1

$$1 + (x')^2 = 1 + \frac{y^2}{9-y^2}$$

$$1 + (x')^2 = \frac{9-y^2}{9-y^2} + \frac{y^2}{9-y^2} = \frac{9}{9-y^2}$$

Lastly take $\sqrt{}$

$$\boxed{\sqrt{1+(x')^2} = \frac{3}{\sqrt{9-y^2}}}$$

$$SA = 2\pi \int_{-1}^{1} 3 \, dy$$

$$SA = 6\pi \left[y \right]_{-1}^{1}$$

$$SA = 6\pi (1 - (-1))$$

$$\boxed{SA = 12\pi \text{ units}^2}$$

1.17 Find the area of the surface generated by revolving the curve $y = \dfrac{1}{3}\left(x^2 + 2\right)^{3/2}$ for $0 \le x \le \sqrt{6}$ about the $y-$axis.

(A) $6\pi\sqrt{6}$ (C) 67π (E) $12\pi\sqrt{6}$ (G) 24π

(B) $3\pi\sqrt{6}$ (D) 12π (F) 36π (H) 6π

1.17 Find the area of the surface generated by revolving the

curve $y = \dfrac{1}{3}\left(x^2 + 2\right)^{3/2}$ for $0 \le x \le \sqrt{6}$ about the $y-$axis.

(A) $6\pi\sqrt{6}$ (C) 67π (E) $12\pi\sqrt{6}$ (G) 24π

(B) $3\pi\sqrt{6}$ (D) 12π (F) 36π (H) 6π

$\left.\begin{array}{l}\text{Surface area}\\ \text{generated by}\\ \text{rotating a curve}\\ \text{about the } y\text{-axis}\end{array}\right\} \quad SA = 2\pi\int x\,ds$

$\left.\begin{array}{l}\text{we are given } x \text{ bounds}\\ \text{and the function is}\\ \text{given as } y = f(x)\end{array}\right\} \text{Do the integral in } x$

$SA = 2\pi\displaystyle\int_a^b x \cdot \sqrt{1+\left(y'_{(x)}\right)^2} \cdot dx = 2\pi\displaystyle\int_0^{\sqrt{6}} x\cdot\boxed{(x^2+1)}\,dx$

$y = \frac{1}{3}(x^2+2)^{3/2}$ $SA = 2\pi\displaystyle\int_0^{\sqrt{6}}(x^3+x)\,dx$

<u>Find y'</u>

$y' = \frac{1}{\cancel{3}}\cdot\frac{\cancel{3}}{2}(x^2+2)^{1/2}\cdot\underbrace{2x}_{\text{chain rule}} = x\sqrt{x^2+2}$

$SA = 2\pi\left[\frac{x^4}{4}+\frac{x^2}{2}\right]_0^{\sqrt{6}}$

<u>Now square</u>

$(y')^2 = \left[x(x^2+2)^{1/2}\right]^2 = x^2(x^2+2)$

$SA = 2\pi\left[\left(\frac{36}{4}+\frac{6}{2}\right)-0\right]$

$(y')^2 = x^4 + 2x^2$

$SA = 2\pi(9+3)$

<u>Now add 1</u>

$1+(y')^2 = x^4 + 2x^2 + 1$

$\boxed{SA = 24\pi \text{ units}^2}$

$1+(y')^2 = (x^2+1)(x^2+1) = (x^2+1)^2$

<u>Finally take $\sqrt{\ }$</u>

$\boxed{\sqrt{1+(y')^2} = x^2+1}$

1.18 Find the surface area generated by revolving the curve $y = \sqrt{3x+1}$ for $1 \le x \le 3$ about the x-axis.

(A) $\dfrac{74\pi}{5}$

(B) $\dfrac{115\pi}{9}$

(C) $\dfrac{97\pi}{12}$

(D) $\dfrac{85\pi}{6}$

(E) $\dfrac{124\pi}{15}$

(F) $\dfrac{109\pi}{9}$

(G) $\dfrac{224\pi}{15}$

(H) $\dfrac{152\pi}{5}$

1.18 Find the surface area generated by revolving the curve $y = \sqrt{3x+1}$ for $1 \leq x \leq 3$ about the $x-$axis.

(A) $\dfrac{74\pi}{5}$

(C) $\dfrac{97\pi}{12}$

(E) $\dfrac{124\pi}{15}$

(G) $\dfrac{224\pi}{15}$

(B) $\dfrac{115\pi}{9}$

(D) $\dfrac{85\pi}{6}$

(F) $\dfrac{109\pi}{9}$

(H) $\dfrac{152\pi}{5}$

Surface area generated
be revolving a curve
about the x-axis
$\Biggr\} \quad SA = 2\pi \displaystyle\int_a^b y\,ds$

we are given x-bounds
and the function is
given as a $y = f(x)$
$\Biggr\} \quad$ Do the integral in X

$$SA = 2\pi \int_1^3 \underbrace{\sqrt{3x+1}}_{y}\sqrt{1+(y'(x))^2}\,dx = 2\pi\int_1^3 \sqrt{3x+1}\cdot\boxed{\frac{\sqrt{12x+13}}{2\sqrt{3x+1}}}\,dx$$

$y = \sqrt{3x+1}$

Find y'

$y' = \dfrac{1}{2\sqrt{3x+1}}\cdot 3 = \dfrac{3}{2\sqrt{3x+1}}$

Now square y'

$(y')^2 = \dfrac{9}{4(3x+1)} = \dfrac{9}{12x+4}$

Next add 1

$1+(y')^2 = 1+\dfrac{9}{12x+4} = \dfrac{12x+4+9}{12x+4}$

$1+(y')^2 = \dfrac{12x+13}{12x+4} = \dfrac{12x+13}{4(3x+1)}$

Finally, take $\sqrt{}$

$$\boxed{\sqrt{1+(y')^2} = \frac{\sqrt{12x+13}}{2\sqrt{3x+1}}}$$

$SA = \pi \displaystyle\int_1^3 \sqrt{12x+13}\,dx$

$u = 12x+13$
$du = 12\,dx$
$\dfrac{du}{12} = \dfrac{12\,dx}{12}$
$\dfrac{1}{12}du = dx$

$\dfrac{1}{12}\int \sqrt{u}\,du$

$\dfrac{1}{12}\cdot \dfrac{u^{3/2}}{\frac{3}{2}} = \dfrac{1}{18}u^{3/2}$

$SA = \dfrac{\pi}{18}\left[(12x+13)^{3/2}\right]_1^3$

$SA = \dfrac{\pi}{18}\left[49^{3/2} - 25^{3/2}\right]$

$SA = \dfrac{\pi}{18}\left[7^3 - 5^3\right]$

$SA = \dfrac{\pi}{18}\left[343 - 125\right]$

$SA = \dfrac{\pi}{18}\left(\overset{109}{\cancel{218}}\right)$

$\begin{array}{r} 49 \\ \times 7 \\ \hline 343 \\ \end{array}$

$\begin{array}{r} 343 \\ -125 \\ \hline 218 \end{array}$

$$\boxed{SA = \frac{109\pi}{9}\ \text{units}^2}$$

1.19 Find the y coordinate of the center of mass of the lamina with constant density ρ bounded above by the graph of $y = 4 - x^2$ and bounded below by the graph of $y = x + 2$ given the fact that the total mass $M = \dfrac{9\rho}{2}$.

(A) $\dfrac{-1}{2}$

(C) $\dfrac{16}{3}$

(E) $\dfrac{12}{5}$

(G) $\dfrac{32}{15}$

(B) $\dfrac{8}{3}$

(D) $\dfrac{24}{5}$

(F) $\dfrac{22}{3}$

(H) $\dfrac{64}{15}$

1.19 Find the y coordinate of the center of mass of the lamina with constant density ρ bounded above by the graph of $y = 4 - x^2$ and bounded below by the graph of $y = x + 2$ given the fact that the total mass $M = \dfrac{9\rho}{2}$.

(A) $\dfrac{-1}{2}$ (C) $\dfrac{16}{3}$ (E) $\dfrac{12}{5}$ (G) $\dfrac{32}{15}$

(B) $\dfrac{8}{3}$ (D) $\dfrac{24}{5}$ (F) $\dfrac{22}{3}$ (H) $\dfrac{64}{15}$

$y = 4 - x^2$
$y = x + 2$

$\bar{y} = \dfrac{M_x}{M} = \dfrac{\frac{54\rho}{5}}{\frac{9\rho}{2}} = \frac{54\rho}{5} \cdot \frac{2}{9\rho}$

$4 - x^2 = x + 2$
$0 = x^2 + x - 2$
$0 = (x+2)(x-1)$
$x = -2 \quad x = 1$

$\boxed{\bar{y} = \dfrac{12}{5}}$

upper   lower

$M_x = \frac{1}{2} \int_a^b \rho \cdot \left[[f(x)]^2 - [g(x)]^2 \right] dx$

$M_x = \frac{\rho}{2} \int_{-2}^{1} \left[(4-x^2)^2 - (x+2)^2 \right] dx$

$M_x = \frac{\rho}{2} \int_{-2}^{1} \left[16 - 8x^2 + x^4 - (x^2 + 4x + 4) \right] dx$

$M_x = \frac{\rho}{2} \int_{-2}^{1} (x^4 - 9x^2 - 4x + 12) dx$

$= \frac{\rho}{2} \left[\frac{x^5}{5} - 3x^3 - 2x^2 + 12x \right]_{-2}^{1}$

$= \frac{\rho}{2} \left[(\frac{1}{5} - 3 - 2 + 12) - (-\frac{32}{5} + 24 - 8 - 24) \right]$

$= \frac{\rho}{2} \left[\frac{1}{5} + 7 + \frac{32}{5} + 8 \right]$

$M_x = \frac{\rho}{2} \left[\frac{33}{5} + 15 \right] = \frac{33 + 75}{5} \cdot \frac{\rho}{2} = \frac{108\rho}{10}$

$\boxed{M_x = \dfrac{54}{5}\rho}$

50

1.20 Find the x coordinate of the centroid (center of mass) of the triangular region with vertices $(0,0)$, $(0,4)$, and $(6,0)$.

(A) 1

(C) 3

(E) $\dfrac{5}{2}$

(G) $\dfrac{7}{4}$

(B) 2

(D) $\dfrac{4}{3}$

(F) $\dfrac{9}{4}$

(H) $\dfrac{3}{2}$

1.20 Find the x coordinate of the centroid (center of mass) of the triangular region with vertices $(0,0)$, $(0,4)$, and $(6,0)$.

(A) 1   (C) 3   (E) $\dfrac{5}{2}$ (G) $\dfrac{7}{4}$

(B) 2   (D) $\dfrac{4}{3}$ (F) $\dfrac{9}{4}$ (H) $\dfrac{3}{2}$

Assume constant density ρ

$$\text{Mass} = \int_a^b \rho \cdot f(x)\, dx$$

$$\text{Mass} = \rho \int_0^6 \left(-\tfrac{2}{3}x + 4\right) dx$$

$\underbrace{\hspace{3cm}}_{\substack{\text{area of the}\\ \text{triangle}}}$

$$\text{Mass} = \rho \cdot \tfrac{1}{2}(\text{base})(\text{height})$$

$$\text{Mass} = \tfrac{\rho}{2}(6)(4) = 12\rho$$

$y = -\tfrac{2}{3}x + 4$

$$\bar{x} = \frac{M_y}{\text{Mass}}$$

$$M_y = \rho \int_a^b x\, f(x)\, dx = \rho \int_0^6 x\left(-\tfrac{2}{3}x + 4\right) dx = \rho \int_0^6 \left(-\tfrac{2}{3}x^2 + 4x\right) dx$$

$$M_y = \rho \left[-\tfrac{2}{3}\cdot\tfrac{x^3}{3} + 4\tfrac{x^2}{2}\right]_0^6 = \rho\left(-\tfrac{2}{9}x^3 + 2x^2\right)\Big|_0^6$$

$$M_y = \rho\left[-\tfrac{2}{9}(216) + 2\cdot 36\right] = \rho\left[-2(24) + 2\cdot 36\right]$$

$$M_y = \rho[-48 + 72] = 24\rho$$

$$\bar{x} = \frac{24\rho}{12\rho} \qquad \boxed{\bar{x} = 2}$$

1.21 Consider the region bounded by $y = x, y = \sin x,$ and $x = \pi$.

If the density is given by $\rho(x) = 2x,$ find the moment about the $y - $ axis.

1.21 Consider the region bounded by $y = x$, $y = \sin x$, and $x = \pi$.

If the density is given by $\rho(x) = 2x$, find the moment about the y-axis.

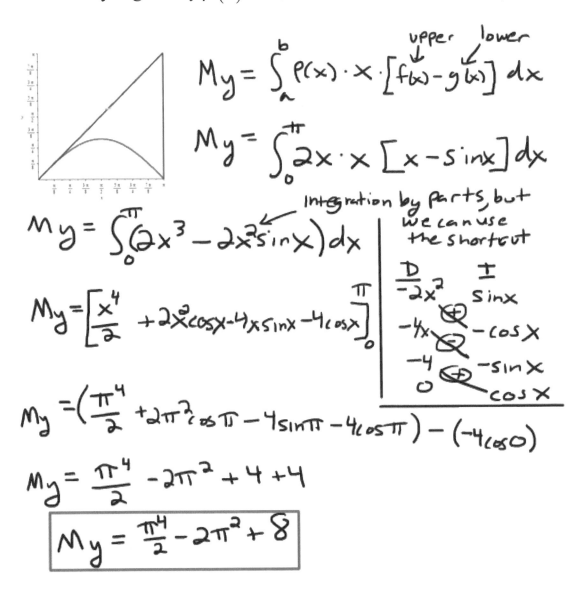

$$M_y = \int_a^b \rho(x) \cdot x \cdot \left[\overset{\text{upper}}{f(x)} - \overset{\text{lower}}{g(x)} \right] dx$$

$$M_y = \int_0^\pi 2x \cdot x \left[x - \sin x \right] dx$$

$$M_y = \int_0^\pi (2x^3 - 2x^2 \sin x) dx$$

Integration by parts, but we can use the shortcut

D	I
$-2x^2$	$\sin x$
$-4x$	$-\cos x$
-4	$-\sin x$
0	$\cos x$

$$M_y = \left[\frac{x^4}{2} + 2x^2\cos x - 4x\sin x - 4\cos x \right]_0^\pi$$

$$M_y = \left(\frac{\pi^4}{2} + 2\pi^2 \cos \pi - 4\sin \pi - 4\cos \pi \right) - \left(-4\cos 0 \right)$$

$$M_y = \frac{\pi^4}{2} - 2\pi^2 + 4 + 4$$

$$\boxed{M_y = \frac{\pi^4}{2} - 2\pi^2 + 8}$$

Section 2: Integration Techniques

Integration By Parts

$$\int u\,dv = u \cdot v - \int v\,du$$

Hierarchy Mneumonic to aid in choosing u

L : logarithmic functions

I : inverse trigonometric functions

A : algebraic functions

T : trigonometric functions

E : exponential functions

(T and E are interchangeable)

Big Picture: We are trading in one integral for another

$$\int u\,dv \qquad \int v\,du$$

Goal : To get a **simpler** integral than the original one

1. Choose u to be a function that becomes simpler when **differentiated**
2. Make sure dv can be readily **integrated**

Shortcut: Works when you have one of the following two situations :
1. $\int (\text{polynomial})(\text{exponential})\,dx$
2. $\int (\text{polynomial})(\text{trig.})\,dx$

$$\int x^2 e^{5x}\,dx$$

Step 1 : Differentiate the polynomial down to 0.

Step 2 : Integrate the trig. or exponential the same amount of times

Step 3 : Multiply along diagonals going down to the right applying an alternating sign starting with +

Diff	Int
$+\ x^2$	e^{5x}
$-\ 2x$	$\frac{1}{5}e^{5x}$
$+\ 2$	$\frac{1}{25}e^{5x}$
0	$\frac{1}{125}e^{5x}$

$$\int x^2 e^{5x}\,dx = \tfrac{1}{5}x^2 e^{5x} - \tfrac{2}{25}xe^{5x} + \tfrac{2}{125}e^{5x} + C$$

Integrating Powers of Trig. Functions

1. $\int \cos^m x \sin^n x \ dx \quad (m, n \text{ positive integers})$

2. $\int \tan^m x \sec^n x \ dx \quad (m, n \text{ positive integers})$

3. $\int \sin(mx)\sin(nx)\,dx \quad \int \cos(mx)\cos(nx)\,dx \quad \int \sin(mx)\cos(nx)\,dx$

$$(m, n \text{ rational with } m \neq n)$$

1. $\int \cos^m x \sin^n x \, dx \quad (m, n \text{ positive integers})$

    A) m, n : **one (or both) odd**   (greater than 1)

        1. Factor out one power from the trig. function that has the odd power, if both have the odd power, just pick one of them and factor out one power.

        2. Use $\cos^2 x + \sin^2 x = 1$ to transform the remaining even power of the above trig function into the other trig. function

        3. Use u – substitution to finish the problem $\left(\text{let } u = \text{"other" trig function} \right)$

    B) m, n : **both even**

        Replace all even powers using the half-angle identities:
        $\sin^2 x = \frac{1}{2}(1 - \cos 2x)$ and $\cos^2 x = \frac{1}{2}(1 + \cos 2x)$

    C) m, n : **one or both** $= 1$

        Use u – substitution. Let $u =$ the trig function with power $\neq 1$
        If both $= 1$, choose either one to be u

2. $\int \tan^m x \sec^n x \, dx \quad (m, n \text{ positive integers})$

    A) m (the power of $\tan x$) : **odd** $\boxed{(\text{must have powers of } \sec x)}$

        1. Factor out one power of $\sec x$ and one power of $\tan x$

        2. Use $\tan^2 x = \sec^2 x - 1$ to transform the remaining even power of $\tan x$ to be in terms of $\sec x$.

        3. Use u – substitution to finish the problem $(\text{let } u = \sec x)$

    B) n (the power of $\sec x$) : **even** $\boxed{(\text{don't need powers of } \tan x)}$

        1. Factor out $\sec^2 x$

        2. If $n > 2$, use $\sec^2 x = 1 + \tan^2 x$ to transform the remaining even power of $\sec x$ to be in terms of $\tan x$

        3. Use u – substitution to finish the problem $(\text{let } u = \tan x)$

    C) Both m **odd** and n **even** : Pick either of the above methods.

    D) If any other combination, then there is no set method.

3. $\int \sin(mx)\sin(nx)\,dx$

$\int \cos(mx)\cos(nx)\,dx$ $\qquad$ $(m,n$ rational with $m \neq n)$

$\int \sin(mx)\cos(nx)\,dx$

We change the product into a sum using the following identities:

$$\sin(mx)\sin(nx) = \frac{1}{2}\Big[\cos\big([m-n]x\big) - \cos\big([m+n]x\big)\Big]$$

$$\cos(mx)\cos(nx) = \frac{1}{2}\Big[\cos\big([m-n]x\big) + \cos\big([m+n]x\big)\Big]$$

$$\sin(mx)\cos(nx) = \frac{1}{2}\Big[\sin\big([m-n]x\big) + \sin\big([m+n]x\big)\Big]$$

Trig. Substitution

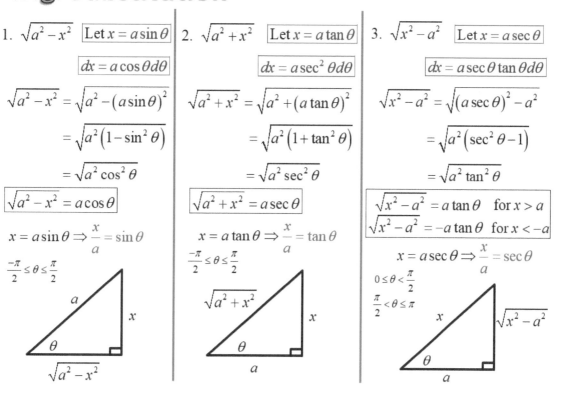

1. $\sqrt{a^2 - x^2}$ $\boxed{\text{Let } x = a\sin\theta}$

$\boxed{dx = a\cos\theta\,d\theta}$

$\sqrt{a^2 - x^2} = \sqrt{a^2 - (a\sin\theta)^2}$

$= \sqrt{a^2(1 - \sin^2\theta)}$

$= \sqrt{a^2 \cos^2\theta}$

$\boxed{\sqrt{a^2 - x^2} = a\cos\theta}$

$x = a\sin\theta \Rightarrow \dfrac{x}{a} = \sin\theta$

$\dfrac{-\pi}{2} \le \theta \le \dfrac{\pi}{2}$

2. $\sqrt{a^2 + x^2}$ $\boxed{\text{Let } x = a\tan\theta}$

$\boxed{dx = a\sec^2\theta\,d\theta}$

$\sqrt{a^2 + x^2} = \sqrt{a^2 + (a\tan\theta)^2}$

$= \sqrt{a^2(1 + \tan^2\theta)}$

$= \sqrt{a^2 \sec^2\theta}$

$\boxed{\sqrt{a^2 + x^2} = a\sec\theta}$

$x = a\tan\theta \Rightarrow \dfrac{x}{a} = \tan\theta$

$\dfrac{-\pi}{2} \le \theta \le \dfrac{\pi}{2}$

3. $\sqrt{x^2 - a^2}$ $\boxed{\text{Let } x = a\sec\theta}$

$\boxed{dx = a\sec\theta\tan\theta\,d\theta}$

$\sqrt{x^2 - a^2} = \sqrt{(a\sec\theta)^2 - a^2}$

$= \sqrt{a^2(\sec^2\theta - 1)}$

$= \sqrt{a^2 \tan^2\theta}$

$\boxed{\begin{array}{l}\sqrt{x^2 - a^2} = a\tan\theta \quad \text{for } x > a \\ \sqrt{x^2 - a^2} = -a\tan\theta \quad \text{for } x < -a\end{array}}$

$x = a\sec\theta \Rightarrow \dfrac{x}{a} = \sec\theta$

$0 \le \theta < \dfrac{\pi}{2}$

$\dfrac{\pi}{2} < \theta \le \pi$

Partial Fraction Decomposition

1. The degree of the denominator **must** be greater than the degree of the numerator $\dfrac{p(x)}{q(x)}$

   If it is not, then **long divide** the denominator into the numerator.

2. Decompose the fraction in the following manner: ($A, B, C,$ and D are constants)

   i) $q(x)$ can be written as a product of only linear polynomials

   $$\frac{5x}{(x-4)(2x+3)} = \frac{A}{x-4} + \frac{B}{2x+3}$$

   ii) $q(x)$ can be written as a product involving powers of linear polynomials

   $$\frac{x^2+6x-4}{(x-3)^3(x+5)} = \frac{A}{x-3} + \frac{B}{(x-3)^2} + \frac{C}{(x-3)^3} + \frac{D}{x+5}$$

   iii) $q(x)$ can be written as a product involving irreducible quadratic polynomials

   $$\frac{16x-5}{(x^2+2x+10)(x-7)} = \frac{Ax+B}{x^2+2x+10} + \frac{C}{x-7}$$

3. Use algebra to find the constants and then integrate the simpler fractions.

$$\int_0^2 \frac{x-12}{x^2+3x-18}dx = \int_0^2 \frac{x-12}{(x+6)(x-3)}\boxed{dx} = \int_0^2 \left(\frac{A}{x+6} + \frac{B}{x-3}\right)dx$$

$$\underbrace{}_{\text{(all linear)}} = \int_0^2 \left(\frac{2}{x+6} + \frac{-1}{x-3}\right)dx$$

$$\frac{x-12}{(x+6)(x-3)} = \frac{A(x-3)}{(x+6)(x-3)} + \frac{B(x+6)}{(x-3)(x+6)}$$

$$x-12 = A(x-3) + B(x+6) \quad \text{true for all } \underline{x} \quad \substack{\text{Choose } x's \\ \text{to make} \\ \text{each term } 0}$$

$$\boxed{\substack{\text{let} \\ x=3}} \quad 3-12 = B(3+6) \Rightarrow -9 = 9B \quad \boxed{B=-1}$$

$$\boxed{\substack{\text{let} \\ x=-6}} \quad -6-12 = A(-6-3) \Rightarrow -18 = -9A \quad A=2$$

$$= \int_0^2 \left(\frac{2}{x+6} + \frac{-1}{x-3}\right)dx = \left(2\cdot\ln|x+6| - \ln|x-3|\right)\Big|_0^2$$

$$\ln(x+6)^2 - \ln|x-3|$$

$$= \ln\left|\frac{(x+6)^2}{x-3}\right|\Big|_0^2 = \ln 64 - \ln 12 = \ln\frac{64}{12} = \boxed{\ln\frac{16}{3}}$$

$$\int \frac{1}{x-4}\,dx = \ln|x-4| + C \qquad\qquad u = x-4 \qquad \int \frac{1}{u}\,du = \ln|u| + C$$
$$du = dx$$

$$\int \frac{1}{(x-4)^2}\,dx = \frac{-1}{x-4} + C \qquad\qquad u = x-4 \qquad \int \frac{1}{u^2}\,du = \int u^{-2}\,du = \frac{-1}{u} + C$$
$$du = dx$$

$$\int \frac{1}{x^2+4}\,dx = \frac{1}{2}\arctan\left(\frac{x}{2}\right) + C \qquad\qquad \frac{1}{x^2+4} = \frac{1}{4\left(\frac{x^2}{4}+1\right)} = \frac{1}{4}\cdot\frac{1}{\left(\frac{x}{2}\right)^2+1}$$

$$\int \frac{1}{x^2+4}\,dx = \frac{1}{4}\int \frac{1}{\left(\frac{x}{2}\right)^2+1}\,dx \qquad\qquad u = \frac{x}{2} \qquad \Rightarrow 2du = dx$$
$$du = \frac{1}{2}dx$$

$$\boxed{\int \frac{1}{x^2+a^2}\,dx = \frac{1}{a}\arctan\left(\frac{x}{a}\right) + C}$$

$$= \frac{1}{4}\int \frac{2}{u^2+1}\,du = \frac{1}{2}\arctan u + C$$

$$\int \frac{x^3-2x^2+18x-29}{x^2+16}\,dx \qquad \text{deg den} < \text{deg num} \quad \maltese \text{ can't start Part.Frac.Decomp.}$$

Must Long Divide.

$$\begin{array}{r} x-2 \\ x^2+0x+16\overline{)x^3-2x^2+18x-29} \\ -(x^3+0x^2+16x) \\ \hline -2x^2+2x-29 \\ -(-2x^2+0x-32) \end{array}$$

hold spot of missing terms

$$x^2 \cdot \boxed{x} = x^3$$

$$x^2 \cdot \boxed{-2} = -2x^2$$

$$\boxed{2x+3}\,\text{rem} \quad \text{Stop}$$

since $2x+3 < x^2+16$ deg1 < deg 2

$$\boxed{\frac{x^3-2x^2+18x-29}{x^2+16} = x-2 + \frac{2x+3}{x^2+16}}$$

$$= x-2 + \frac{2x}{x^2+16} + \frac{3}{x^2+16}$$

u-sub   formula

$$\begin{array}{r} 205 \\ 6\overline{)12304} \\ -12 \\ \hline 03 \\ 0 \\ \hline 34 \\ -30 \\ \hline \textcircled{4}\ \text{remainder} \end{array}$$

$$\frac{12304}{6} = 205\frac{4}{6}$$

$$\underline{A}\,+\frac{B}{}$$

2.1 Evaluate the integral

$$\int_0^4 x^2 e^{\frac{x}{2}}dx$$

(A) $\frac{1}{3}\left(\sqrt{e}-1\right)$ (C) $8\left(e^2-2\right)$ (E) $16\left(e^2-1\right)$

(B) $\sqrt{e}+1$ (D) $16\left(e^2-10\right)$ (F) $8e-16$

$u = x^2$
$du = 2x$

$dv = e^{\frac{1}{2}x}$
$v = 2e^{\frac{1}{2}x}$

$2x^2 e^{\frac{1}{2}x} - \int 2e^{\frac{1}{2}x} \cdot 2x$

$2x^2 e^{\frac{1}{2}x} - 4\int e^{x \cdot \frac{1}{2}x}$

2.1 Evaluate the integral

$$\int_0^4 x^2 e^{\frac{x}{2}} dx$$

(A) $\frac{1}{3}\left(\sqrt{e}-1\right)$ (C) $8\left(e^2-2\right)$ (E) $16\left(e^2-1\right)$

(B) $\sqrt{e}+1$ (D) $16\left(e^2-10\right)$ (F) $8e-16$

Integration by parts twice
 or use the shortcut

$$\begin{array}{cc} D & I \\ \frac{}{} & \frac{}{} \\ x^2 & \oplus \quad e^{x/2} \\ 2x & \ominus \quad 2e^{x/2} \\ 2 & \oplus \quad 4e^{x/2} \\ 0 & 8e^{x/2} \end{array} \qquad \int e^{Kx}dx = \frac{1}{K}e^{Kx}+C$$

$$= \left[2x^2 e^{x/2} - 8xe^{x/2} + 16e^{x/2}\right]_0^4$$

$$= 2\left[e^{x/2}\left(x^2 - 4x + 8\right)\right]_0^4$$

$$= 2\left[e^2\left(16-16+8\right) - 1\cdot\left(0-0+8\right)\right]$$

$$= 2\left[8e^2 - 8\right] \quad = \boxed{16\left(e^2-1\right)}$$

2.2 Evaluate the integral

$$\int_0^1 x^2 \arctan(x)\, dx$$

 write down unit circle

(A) $\dfrac{\pi}{4} + \dfrac{1}{12}\ln 4$

(C) $\dfrac{1}{12} + \ln 2$

(E) $\dfrac{\pi}{6} + \dfrac{1}{6}\ln 4$

(B) $\dfrac{\pi}{12} - \dfrac{1}{6} + \dfrac{1}{6}\ln 2$

(D) $\dfrac{\pi}{3} + \dfrac{1}{3}\ln 2$

(F) $\dfrac{\pi}{12} + \dfrac{1}{6}\ln 2$

$$\frac{-1}{2} \cdot \frac{2}{\sqrt{3}} \quad \frac{\ln}{4}$$

$$uv - \int v\, du$$

L
I
A
T
E

$$\frac{\sqrt{3}}{4} \qquad \frac{-\frac{1}{2}}{-\frac{\sqrt{3}}{2}} \qquad \frac{\sqrt{3}}{2} \frac{1}{2} =$$

$$\frac{-1 \cdot 2}{2 \cdot -\sqrt{3}} = \frac{-2}{\sqrt{3}}$$

$$\frac{1}{3}x^3 \arctan x - \int \frac{1}{3}x^3 \left(\frac{1}{1+x^2}\right)$$

$$\frac{1}{3}x^3 \arctan x - \frac{1}{3}$$

$$v = \arctan x \quad dv = x^2$$
$$du = \frac{1}{1+x^2} \qquad v = \frac{1}{3}x^3$$

$$\frac{1}{3}x^3 \arctan x - \frac{x^6}{18}$$

$$v = 1+x^2$$
$$dv = 2x\, dx$$

D	I
arctan x ⊕	x^2
$\frac{1}{1+x^2}$ ⊖	$2x$
⊕	2
⊖	0

$$\frac{1}{3}\arctan(1) - \frac{1}{18}$$

$$\frac{1}{3}\left(\frac{\pi}{4}\right) - \frac{1}{18}$$

$$-\frac{1}{3}\int x^3 \left(\frac{1}{1+x^2}\right)$$

$$\frac{\pi}{12} - \frac{1}{18}$$

$$-\frac{1}{3}\int \frac{x^3}{1+x^2} \qquad v = x^3$$
$$dv = 3x^2\, dx$$

2.2 Evaluate the integral

$$\int_0^1 x^2 \arctan(x)\, dx$$

(A) $\dfrac{\pi}{4} + \dfrac{1}{12}\ln 4$

(C) $\dfrac{1}{12} + \ln 2$

(E) $\dfrac{\pi}{6} + \dfrac{1}{6}\ln 4$

(B) $\dfrac{\pi}{12} - \dfrac{1}{6} + \dfrac{1}{6}\ln 2$

(D) $\dfrac{\pi}{3} + \dfrac{1}{3}\ln 2$

(F) $\dfrac{\pi}{12} + \dfrac{1}{6}\ln 2$

Use Integration by Parts
  but you can't use the shortcut

$$u = \arctan x \qquad dv = x^2$$

$$du = \frac{1}{1+x^2}\, dx \qquad v = \frac{x^3}{3}$$

$$uv - \int v\, du$$

$$= \frac{x^3}{3}\arctan x - \frac{1}{3}\int \frac{x^3}{1+x^2}\, dx$$

$$\text{use Long Division}$$

$$\begin{array}{r} x \\ x^2+1\overline{\smash{\big)}\,x^3} \\ \underline{-(x^3+x)} \\ -x \end{array}$$

$$= \frac{x^3}{3}\arctan x - \frac{1}{3}\int \left(x - \frac{x}{x^2+1}\right) dx$$

$$= \frac{x^3}{3}\arctan x - \frac{1}{3}\left[\frac{x^2}{2} - \frac{1}{2}\ln(x^2+1)\right]\Big|_0^1$$

$$= \left(\frac{1}{3}\arctan 1 - \frac{1}{3}\left[\frac{1}{2} - \frac{1}{2}\ln 2\right]\right) - \left(0 - \frac{1}{3}[0-0]\right)$$

$$= \frac{1}{3}\cdot\frac{\pi}{4} - \frac{1}{3}\left(\frac{1}{2} - \frac{1}{2}\ln 2\right)$$

$$= \boxed{\frac{\pi}{12} - \frac{1}{6} + \frac{1}{6}\ln 2}$$

2.3 Evaluate the integral

$$\int_{1}^{\sqrt[3]{e}} 18x^2 \ln(x)\, dx$$

(A) $\dfrac{3}{2}$ (C) $\dfrac{5}{2}$ (E) $\dfrac{7}{2}$ (G) 1

(B) $\dfrac{1}{2}$ (D) 2   (F) 3   (H) 4

LIATE

$U = \ln x \qquad dv = 18x^2$

$dv = \dfrac{1}{x} \qquad\quad v = 6x^3$

$6x^3 \ln x - \int 6x^3 \cdot \dfrac{1}{x}$

$6x^3 \ln x - \int 6x^2 \cdot 3$

$6x^3 \ln x - 2x^3 \Big|_1^{\sqrt[3]{e}}$

$2x^3(3\ln x - 1)$

$2(e)(3\ln(\sqrt[3]{e}) - 1) - 2(3 - 1)$

$+2$

$6(\sqrt[3]{e})\ln(\sqrt[3]{e}^{\frac{1}{3}}) - 2e$

$6e(\frac{1}{3}) - 2e = 0$

$2e - 2e = 0$

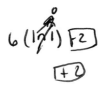

$\boxed{+2}$

2.3 Evaluate the integral

$$\int_{1}^{\sqrt[3]{e}} 18x^2 \ln(x)\, dx$$

(A) $\dfrac{3}{2}$ (C) $\dfrac{5}{2}$ (E) $\dfrac{7}{2}$ (G) 1

(B) $\dfrac{1}{2}$ (D) 2   (F) 3   (H) 4

Use Integration by Parts
but you can't use the shortcut

$u = \ln x$ $dv = 18x^2$

$du = \frac{1}{x}dx$ $v = 6x^3$

$uv - \int v\, du$

$= 6x^3 \cdot \ln x - \int 6x^2 dx$

$= \left[6x^3 \cdot \ln x - 2x^3 \right] \Big|_{1}^{\sqrt[3]{e}}$

$= \left[6\left(\sqrt[3]{e}\right)^3 \cdot \ln \sqrt[3]{e} - 2\left(\sqrt[3]{e}\right)^3 \right] - (0-2)$

$= 6e \cdot \ln e^{1/3} - 2e + 2$

$= 6e \cdot \frac{1}{3} - 2e + 2$

$= 2e - 2e + 2$

$= \boxed{2}$

2.4 Evaluate the integral

$$\int_0^{\pi/6} x\sin(3x)\,dx$$

(A) $\dfrac{1}{2}$ (C) $\dfrac{1}{3}$ (E) $\dfrac{1}{4}$ (G) $\dfrac{3}{4}$

(B) $\dfrac{1}{8}$ (D) $\dfrac{1}{16}$ (F) $\dfrac{1}{9}$ (H) $\dfrac{2}{3}$

write unit circle

a poly (trig)

LIATE

$u = x$ $dv = \sin(3x)$
$du = 1$ $v = \frac{1}{3}\cos(3x)$

$$-\frac{1}{3}x\cos(3x) - \int \frac{1}{3}\cos(3x) \qquad \begin{array}{cc} U & d \\ + x & \sin(3x) \\ - 1 & 3\cos(3x) \\ + 0 & -9\sin(3x) \end{array}$$

$$-\frac{1}{3}x\cos(3x) -$$

$$3x\cos(3x) + 9\sin(3x)$$

$$\frac{3\pi}{6}$$

$$\frac{1}{2}\pi\left(\cos\left(\frac{3\pi}{6}\right) + 9\sin\left(\frac{3\pi}{6}\right)\right)$$

$$\overset{0}{\frac{1}{2}\pi\left(\cos\frac{1}{2}\pi\right) + 9\sin\left(\frac{1}{2}\pi\right)}$$

69

2.4 Evaluate the integral

$$\int_0^{\pi/6} x \sin(3x)\, dx$$

(A) $\dfrac{1}{2}$ (C) $\dfrac{1}{3}$ (E) $\dfrac{1}{4}$ (G) $\dfrac{3}{4}$

(B) $\dfrac{1}{8}$ (D) $\dfrac{1}{16}$ (F) $\dfrac{1}{9}$ (H) $\dfrac{2}{3}$

Use Integration by Parts Shortcut

$$\begin{array}{ccc} \underline{D} & \underline{I} & \int \sin kx\, dx \\ x & \sin(3x) & = -\tfrac{1}{k}\cos kx + c \\ 1 & -\tfrac{1}{3}\cos(3x) & \\ 0 & -\tfrac{1}{9}\sin(3x) & \end{array}$$

$$= \left[-\tfrac{x}{3}\cos(3x) + \tfrac{1}{9}\sin(3x) \right]_0^{\pi/6}$$

$$= \left(-\tfrac{\pi}{18}\cos(\pi/2) + \tfrac{1}{9}\sin(\tfrac{\pi}{2}) \right) - (0 + 0)$$

with underbraces: $\cos(\pi/2)=0$, $\sin(\tfrac{\pi}{2})=1$

$$= \boxed{\tfrac{1}{9}}$$

Anti-deriv. w/o shortcut:

$$u = x \qquad dv = \sin(3x)$$
$$du = dx \qquad v = -\tfrac{1}{3}\cos(3x)$$

$$uv - \int v\, du$$

$$= -\tfrac{1}{3}x \cdot \cos(3x) + \tfrac{1}{3}\int \cos(3x)\, dx$$

$$= -\tfrac{1}{3}x\cos(3x) + \tfrac{1}{3}\cdot\tfrac{1}{3}\sin(3x)$$

$$= -\tfrac{1}{3}x\cos(3x) + \tfrac{1}{9}\sin(3x)$$

2.5 Evaluate the integral

$$\int_0^4 x^2 \cos\left(\frac{\pi}{2}x\right)dx$$

(A) $\dfrac{4}{\pi} - \dfrac{16}{\pi^3}$

(C) $\dfrac{64}{\pi^2}$

(E) $\dfrac{-16}{\pi^2}$

(G) $\dfrac{16}{\pi^2}$

(B) $\dfrac{8}{\pi} + \dfrac{16}{\pi^3}$

(D) $\dfrac{8}{\pi}$

(F) $\dfrac{8}{\pi^2}$

(H) $\dfrac{32}{\pi^2}$

$$\begin{array}{cc}
\mathbf{D} & \mathbf{I} \\
+ \quad x^2 & \cos(\tfrac{\pi}{2}x) \\
- \quad 2x & \tfrac{2}{\pi}\sin(\tfrac{\pi}{2}x) \\
+ \quad 2 & \\
- \quad 0 &
\end{array}$$

$u = \tfrac{1}{2}\pi x$

$\tfrac{\pi}{2} \cdot \frac{2}{\pi} = \frac{\pi}{2}$ $\frac{\frac{1}{2}}{3}\cos(3$

$2\pi \cdot \frac{\pi}{2} = 2\pi^2$ $\frac{2}{\pi}$

2.5 Evaluate the integral

$$\int_0^4 x^2 \cos\left(\frac{\pi}{2}x\right)dx$$

(A) $\dfrac{4}{\pi}-\dfrac{16}{\pi^3}$ (C) $\dfrac{64}{\pi^7}$ (E) $\dfrac{-16}{n^2}$ (G) $\dfrac{16}{\pi^2}$

(B) $\dfrac{8}{\pi}+\dfrac{16}{\pi^3}$ (D) $\dfrac{8}{\pi}$ (F) $\dfrac{8}{\pi^2}$ (H) $\dfrac{32}{\pi^2}$

Use Integration by Parts
(a) with the shortcut
(b) without the shortcut

(a)

$\begin{array}{ll} \boxed{+} & x^2 \\ & \cos\frac{\pi}{2}x \\ \boxed{-}\; 2x & \frac{2}{\pi}\sin\frac{\pi}{2}x \\ \boxed{+}\; 2 & -\frac{4}{\pi^2}\cos\frac{\pi}{2}x \\ \boxed{-}\; 0 & -\frac{8}{\pi^3}\sin(\frac{\pi}{2}x) \end{array}$

$= \frac{2x^2}{\pi}\sin\frac{\pi}{2}x + \frac{8x}{\pi^2}\cos\frac{\pi}{2}x - \frac{16}{\pi^3}\sin\frac{\pi}{2}x$

(b) $u = x^2$ $dv = \cos\frac{\pi}{2}x$
$du = 2x$ $v = \frac{2}{\pi}\sin\frac{\pi}{2}x$

$= uv - \int v\, du$

$= \frac{2x^2}{\pi}\sin\frac{\pi}{2}x - \frac{4}{\pi}\underbrace{\int x\sin\frac{\pi}{2}x\, dx}_{\text{I.B.P.}}$

$u = x$ $dv = \sin\frac{\pi}{2}x$
$du = dx$ $v = -\frac{2}{\pi}\cos\frac{\pi}{2}x$

$\Rightarrow$ $uv - \int v\, du$

$= \frac{2x^2}{\pi}\sin(\frac{\pi}{2}x) - \frac{4}{\pi}\left[-\frac{2x}{\pi}\cos\frac{\pi}{2}x + \frac{2}{\pi}\int\cos\frac{\pi}{2}x\, dx\right]$

$\frac{2}{\pi}(\frac{2}{\pi}\sin\frac{\pi}{2}x)$

$= \frac{2x^2}{\pi}\sin(\frac{\pi}{2}x) - \frac{4}{\pi}\left[\frac{2x}{\pi}\cos(\frac{\pi}{2}x) + \frac{4}{\pi^2}\sin\frac{\pi}{2}x\right]$

$= \frac{2x^2}{\pi}\sin(\frac{\pi}{2}x) + \frac{8x}{\pi^2}\cos(\frac{\pi}{2}x) - \frac{16}{\pi^3}\sin(\frac{\pi}{2}x)$

$= \frac{2}{\pi}\left[x^2\sin\frac{\pi}{2}x + \frac{4}{\pi}x\cos(\frac{\pi}{2}x) - \frac{8}{\pi^2}\sin(\frac{\pi}{2}x)\right]\Big|_0^4$

$= \frac{2}{\pi}\left[(16\sin 2\pi + \frac{16}{\pi}\cos 2\pi - \frac{8}{\pi^2}\sin 2\pi) - (0 - 0 - 0)\right]$

$= \frac{2}{\pi}\cdot\frac{16}{\pi} = \boxed{\dfrac{32}{\pi^2}}$

2.6 Evaluate the integral

$$\int_0^{\pi/2} \sqrt{\cos x}\,\sin^3 x\, dx$$

(A) $\dfrac{11}{20}$ (C) $\dfrac{9}{20}$ (E) $\dfrac{8}{21}$ (G) $\dfrac{2}{5}$

(B) $\dfrac{3}{7}$ (D) $\dfrac{4}{11}$ (F) $\dfrac{1}{2}$ (H) $\dfrac{11}{21}$

2.6 Evaluate the integral

$$\int_0^{\pi/2} \sqrt{\cos x}\, \sin^3 x \, dx$$

(A) $\dfrac{11}{20}$ (C) $\dfrac{9}{20}$ (E) $\dfrac{8}{21}$ (G) $\dfrac{2}{5}$

(B) $\dfrac{3}{7}$ (D) $\dfrac{4}{11}$ (F) $\dfrac{1}{2}$ (H) $\dfrac{11}{21}$

<u>Power of sinx is odd</u>:
- factor out one power of sinx

$$= \int_0^{\pi/2} \sqrt{\cos x}\ \sin^2 x \cdot \sin x \, dx$$

- Transform remaining powers
  using $\sin^2 x = 1 - \cos^2 x$

$$= \int_0^{\pi/2} \sqrt{\cos x}\,(1 - \cos^2 x)\sin x \, dx$$

- Let $u = \cos x$ $x=0 \Rightarrow u=1$ uL

  $du = -\sin x\, dx$ $x=\frac{\pi}{2} \Rightarrow u=0$ uL

  $-1 \cdot du = \sin x\, dx$

$$= \int_1^0 \sqrt{u}\,(1-u^2) \cdot -1 \cdot du \quad \text{LL} > \text{UL}$$

$$= -\int_0^1 (\sqrt{u} - u^{5/2}) \cdot -1\, du$$

$$= \int_0^1 (u^{1/2} - u^{5/2})\, du = \left[\frac{2}{3} u^{3/2} - \frac{2u^{7/2}}{7} \right]_0^1$$

$$= \left(\frac{2}{3} - \frac{2}{7} \right) - 0 = \frac{14-6}{21} = \boxed{\frac{8}{21}}$$

74

2.7 Evaluate the integral

$$\int_0^{\pi/6} \sin^3(3x)\,dx$$

(A) $\dfrac{1}{3}$ (C) $\dfrac{2}{9}$ (E) $\dfrac{1}{9}$ (G) $\dfrac{5}{3}$

(B) $\dfrac{7}{9}$ (D) $\dfrac{4}{9}$ (F) $\dfrac{4}{3}$ (H) $\dfrac{2}{3}$

$$\int_0^{\pi/6} \sin^2(3x)\,\sin(3x)\,dx$$

$$(1 - \cos^2(3x))\,\sin(3x)\,dx$$

$$v = \cos(3x)$$
$$dv = -3\sin(3x)$$
$$-\frac{1}{3}\,dv = \sin 3x\,dx$$

$$\approx \frac{1}{3}\int (1 - v^2)$$

$$\frac{1}{3}\int (v^2 - 1)$$

$$\frac{1}{3}\left(\frac{v^3}{3} - v\right)_0$$

$$\frac{1}{3}\left(\frac{\cos(3x)^3}{3} - \cos(3x)\right)\Big|_0^{\pi/6}$$

$$\cos\left(\frac{v}{2}\right)^3 - \cos\frac{\pi}{6}\cdot\frac{\pi}{2}$$

$$\approx \frac{1}{3}\left(\frac{\cos\frac{\pi}{2}}{3} - \frac{1}{3}\cdot 1 - 1 = -\frac{2}{3}\right)\quad \frac{2}{9}$$

75

2.7 Evaluate the integral

$$\int_{0}^{\pi/6} \sin^3 (3x)\, dx$$

(A) $\dfrac{1}{3}$ (C) $\dfrac{2}{9}$ (E) $\dfrac{1}{9}$ (G) $\dfrac{5}{3}$

(B) $\dfrac{7}{9}$ (D) $\dfrac{4}{9}$ (F) $\dfrac{4}{3}$ (H) $\dfrac{2}{3}$

Power of sin(3x) is odd
 • factor out one power of sin(3x)

$$= \int_{0}^{\pi/6} \sin^2(3x) \cdot \sin(3x)\, dx$$

 • transform remaining powers of sin(3x)
   by $\sin^2(3x) = 1 - \cos^2(3x)$

$$= \int_{0}^{\pi/6} (1 - \cos^2(3x))\, \sin(3x)\, dx$$

 • let $u = \cos(3x)$
       $du = -3\sin(3x)\, dx$
       $-\tfrac{1}{3} du = \sin(3x)\, dx$

$$\int (1-u^2) \cdot \tfrac{-1}{3}\, du = \tfrac{+1}{3} \int (u^2-1)\, du$$
$$= \tfrac{1}{3}\left(\tfrac{u^3}{3} - u \right)$$
$$= \tfrac{1}{3}\left[\tfrac{1}{3}\cos^3(3x) - \cos(3x) \right]_{0}^{\pi/6}$$

$$= \tfrac{1}{3}\left[\left(\tfrac{1}{3}\cos\tfrac{\pi}{2}\right)^3 - \cos\tfrac{\pi}{2}\right) - \left(\tfrac{1}{3} \cdot 1 - 1\right) \right]$$

$$= \tfrac{1}{3}\left[-(-\tfrac{2}{3}) \right] = \boxed{\tfrac{2}{9}}$$

2.8 Evaluate the integral

$$\int_0^\pi 40\sin^4(x)\,dx$$

(A) $\dfrac{\pi}{4}$

(B) $\dfrac{4\pi}{3}$

(C) 3π

(D) 15π

(E) 5

(F) 4π

(G) $\dfrac{5\pi}{3}$

(H) 45π

$A \int_0^{\wedge} u$

$40 \int_0^{\wedge} \cdot \frac{1}{2}(1-\cos2x)\sin^2(x)dx$

$1-\cos2x$

$-\cos2x$

$-\cos2x -\cos2x \quad \cos^2$

$\begin{matrix}\cos2x\\\cos2x\end{matrix}$

$20\int_0^{\wedge} (1-\cos(2x))(\sin^2(x))dx$

$40 \int_{\cup}^{\wedge} \frac{1}{2}(1-\cos2x)\frac{1}{2}(1-\cos2x)$

$u=\cos(2x)$

$du=-2\sin(2x)$

$10\int_0^{\wedge} (1-\cos2x)(1-\cos2x)$

2.8 Evaluate the integral

$$\int_0^\pi 40 \sin^4(x)\, dx$$

(A) $\dfrac{\pi}{4}$ (C) 3π (E) 5 (G) $\dfrac{5\pi}{3}$

(B) $\dfrac{4\pi}{3}$ (D) 15π (F) 4π (H) 45π

Only power present even:

• use $\sin^2 x = \frac{1}{2}(1-\cos 2x)$ $\cos^2 x = \frac{1}{2}(1+\cos 2x)$

to transform all even powers of sinx

$\sin^4 x = \sin^2 x \cdot \sin^2 x$

$\sin^4 x = \frac{1}{2}(1-\cos 2x)\frac{1}{2}(1-\cos 2x)$

$\sin^4 x = \frac{1}{4}(1 - 2\cos 2x + \cos^2 2x)$

$\sin^4 x = \frac{1}{4}\left(1 - 2\cos 2x + \frac{1}{2}(1+\cos 4x)\right)$

$\sin^4 x = \frac{1}{4}\left(\frac{1}{2} - 2\cos 2x + \frac{1}{2} + \frac{1}{2}\cos 4x\right)$

$\sin^4 x = \frac{1}{4}\left(\frac{3}{2} - 2\cos 2x + \frac{1}{2}\cos 4x\right)$

$= 40 \int_0^\pi \frac{1}{4}\left(\frac{3}{2} - 2\cos 2x + \frac{1}{2}\cos 4x\right) dx$

$= 10 \left[\int_0^\pi \frac{3}{2}\, dx - \underbrace{\int_0^\pi 2\cos 2x\, dx}_{\substack{0 \text{ since it} \\ \text{is a full} \\ \text{period of} \\ \cos 2x}} + \underbrace{\int_0^\pi \frac{1}{2}\cos 4x\, dx}_{\substack{0 \text{ since it} \\ \text{is 2 full periods} \\ \text{of } \cos 4x}} \right]$

$= 15\int_0^\pi dx = 15[x]_0^\pi = \boxed{15\pi}$

78

2.9 Evaluate the integral

$$\int_0^{\pi/3} \tan^3(x)\sec^3(x)\ dx$$

(A) $\dfrac{58}{15}$ (C) $\dfrac{7}{3}$ (E) $\dfrac{14}{3}$ (G) $\dfrac{46}{15}$

(B) $\dfrac{31}{5}$ (D) $\dfrac{52}{5}$ (F) $\dfrac{62}{5}$ (H) $\dfrac{93}{12}$

$$\int_0^{\pi/3} \tan x \sec x \ \tan^2(x)\sec^2(x)$$

$$\int_0^{\pi/3} \tan x \sec x \ \sec^2 x - 1 \ \sec^2 x$$

$v = \sec x$

$dv = \sec x \tan x \ dx$

$$\int_0^{\pi/3} v^2(v^2-1)$$

$\sec = \dfrac{1}{\cos x}$

$$\int_0^{\pi/3} v^4 - v^2$$

2

$$\left. \dfrac{v^5}{4\cdot5} - \dfrac{v^3}{3} \right|_0^{\pi/3}$$

$2 \cdot 2 \cdot 2 \cdot 2$

$$\dfrac{2^5}{4\cdot5} - \dfrac{2^3}{2\cdot3} - \emptyset$$

$$\dfrac{16}{4} - \dfrac{4}{2} = 4 - 2 = 2$$

2.9 Evaluate the integral

$$\int\limits_{0}^{\pi/3} \tan^3(x)\sec^3(x)\ dx$$

(A) $\dfrac{58}{15}$ (C) $\dfrac{7}{3}$ (E) $\dfrac{14}{3}$ (G) $\dfrac{46}{15}$

(B) $\dfrac{31}{5}$ (D) $\dfrac{52}{5}$ (F) $\dfrac{62}{5}$ (H) $\dfrac{93}{12}$

<u>Power of tanx odd with secx powers present:</u>

• Factor out tanx·secx

$$= \int_{0}^{\pi/3} \tan^2 x \cdot \sec^2 x \cdot \tan x \sec x\, dx$$

• Transform remaining even power
of tanx using
$$\tan^2 x = \sec^2 x - 1$$

$$= \int_{0}^{\pi/3} (\sec^2 x - 1)\sec^2 x\, \tan x \sec x\, dx$$

• Let $u = \sec x$
$$du = \sec x \tan x\, dx$$

$$= \int (u^2 - 1) u^2 \cdot du = \int (u^4 - u^2)\, du$$

$$= \frac{u^5}{5} - \frac{u^3}{3} \Rightarrow \left[\frac{1}{5}(\sec x)^5 - \frac{1}{3}(\sec x)^3 \right]_{0}^{\pi/3}$$

$$\sec \frac{\pi}{3} = \frac{1}{\cos \frac{\pi}{3}} = \frac{1}{\frac{1}{2}} = 2 \qquad \sec 0 = \frac{1}{\cos 0} = 1$$

$$= \left(\frac{1}{5}(2^5) - \frac{1}{3}(2^3) \right) - \left(\frac{1}{5} - \frac{1}{3} \right)$$

$$= \frac{32}{5} - \frac{8}{3} - \frac{1}{5} + \frac{1}{3} = \frac{31}{5} - \frac{7}{3}$$

$$= \frac{93 - 35}{15} = \boxed{\frac{58}{15}}$$

2.10 Evaluate the integral

$$\int_0^{\pi/3} 15\sec^4(x)\tan^2(x)\,dx$$

(A) $42\sqrt{3}$ (C) $36\sqrt{3}$ (E) $45\sqrt{3}$ (G) $15\sqrt{3}$

(B) $\dfrac{4\pi}{3}$ (D) $27\sqrt{3}$ (F) $12\sqrt{3}$ (H) $30\sqrt{3}$

$\frac{1}{2}+\frac{1}{2}$

$\sqrt{3}\cdot\sqrt{3}=\pm 3$

$(\sqrt{3})^2 \qquad \frac{1}{2}\cdot 2$

$$15\int_0^{\pi/3} \sec^2(x)\,\sec^2(x)\,\tan^2(x)\,dx$$

$u = \tan x$

$du = \sec^2 x$

$\dfrac{81}{3}$

$9 \;\; 27 \;\; 81 \qquad \dfrac{2^{40}s}{3}\,\boxed{243}$

$\overset{\frown}{3\cdot3\cdot3\cdot3\cdot3}\;\dfrac{}{\sqrt{3}}$

$$15\int_0^{\pi/3} (1+\tan^2 x)(1+\tan^2(x))\tan^2 x\,dx$$

$.3=9$

$\overset{3}{\overbrace{\sqrt{3}\cdot\sqrt{3}}}\cdot\overset{}{\overbrace{\sqrt{3}\cdot\sqrt{3}\cdot\sqrt{3}}}$

$$15\int_0^{\pi/3}(1+v^2)(v^2)\,dx$$

$8(\sqrt{3})^5 + 5(\sqrt{3})^3$

27

$3\,\sqrt{3}(3)^{\frac{5}{2}} + 5(3)^{\frac{3}{2}}$

$\dfrac{\sin}{\cos}$

$\overset{3}{5\sqrt{3}\sqrt{3}\,\sqrt{3}}$

$\dfrac{15}{\boxed{42\sqrt{3}}}$

$$15\int_0^{\pi/3} v^2 + v^4$$

$\sqrt{3}$

$$15\int \frac{v^5}{5}+\frac{v^3}{3} - \frac{v^5}{5}+\frac{v^3}{3}$$

$\dfrac{\frac{\sqrt{3}}{2}}{\frac{1}{2}} \qquad \dfrac{2\sqrt{3}}{2}=\sqrt{3}$

$$15\int \frac{(\tan(x))^5}{5}+\frac{(\tan x)^3}{3} - \frac{(\tan x)^5}{5}+\frac{(\tan x)^3}{3}$$

$\dfrac{0}{1}=0$

$\dfrac{(\sqrt{3})^5}{5}+\dfrac{(\sqrt{3})^3}{3} - \cancel{0}+0$

81

2.10 Evaluate the integral

$$\int_{0}^{\pi/3} 15\sec^4(x)\tan^2(x)\,dx$$

(A) $42\sqrt{3}$ (C) $36\sqrt{3}$ (E) $45\sqrt{3}$ (G) $15\sqrt{3}$

(B) $\dfrac{4\pi}{3}$ (D) $27\sqrt{3}$ (F) $12\sqrt{3}$ (H) $30\sqrt{3}$

Power of secx even:

- Factor out $\sec^2 x$

$$= \int_0^{\pi/3} 15 \cdot \sec^2 x \cdot \tan^2 x \cdot \sec^2 x\, dx$$

- Transform remaining even powers of secx
  by $\sec^2 x = 1 + \tan^2 x$

$$= \int_0^{\pi/3} 15 \cdot (1 + \tan^2 x)\,\tan^2 x \,\sec^2 x\, dx$$

- Let $u = \tan x$ ul $x = 0$ $u = 0$
  $du = \sec^2 x\, dx$ ul $x = \frac{\pi}{3}$ $u = \sqrt{3}$

$$= \int_0^{\sqrt{3}} 15 \cdot (1 + u^2)\,u^2 \cdot du = 15\int_0^{\sqrt{3}} (u^2 + u^4)\,du$$

$$= 15 \cdot \left(\frac{u^3}{3} + \frac{u^5}{5} \right)_0^{\sqrt{3}} = \left(5u^3 + 3u^5 \right)_0^{\sqrt{3}}$$

$$= 5(\sqrt{3})^3 + 3(\sqrt{3})^5 = \underbrace{5 \cdot 3\sqrt{3}}_{15} + \underbrace{3 \cdot 9\sqrt{3}}_{27}$$

$$= \boxed{42\sqrt{3}}$$

2.11 Evaluate the integral

$$\int_0^{\pi/12} \sin(3x)\sin x \, dx$$

(A) $\dfrac{3+\sqrt{3}}{12}$ (C) $\dfrac{4-\sqrt{3}}{3}$ (E) $\dfrac{4-\sqrt{2}}{8}$ (G) $\dfrac{4+\sqrt{2}}{8}$

(B) $\dfrac{2+\sqrt{2}}{4}$ (D) $\dfrac{3+\sqrt{2}}{12}$ (F) $\dfrac{2-\sqrt{3}}{16}$ (H) $\dfrac{2+\sqrt{3}}{16}$

(F is circled)

$$\int_0^{\pi/12} \frac{1}{2}\left[\cos[3-1]x) - \cos[3+1]x)\right]$$

$$\int_0^{\pi/12} \frac{1}{2}\cos 2x - \frac{1}{2}\cos 4x \, dx$$

$$\left\{ \; \frac{1}{2}\int_0^{\pi/12}\cos 2x - \frac{1}{2}\int_0^{\pi/12}\cos 4x \, dx \right.$$

$$u = 2x$$
$$du = 2dx$$
$$\frac{1}{2}du = dx$$
$$\frac{1}{2}u.$$
$$\frac{1}{2}$$

$$\frac{1}{4}\Big|\frac{1}{2}\sin(2x) - \frac{1}{2}\Big|\frac{1}{24}\sin 4x \, dx$$

$$\frac{1}{4}\sin(2x) - \frac{1}{8}\sin 4x \, dx$$

$$\frac{1}{4}\sin\left(\frac{4}{6}\right) - \frac{1}{8}\sin\left(\frac{4}{3}\right)dx \qquad -\int 0$$

$$\frac{1}{4}\left(\frac{1}{2}\right) - \frac{1}{8}\left(\frac{\sqrt{3}}{2}\right)dx$$

$$\frac{1}{8} - \frac{\sqrt{3}}{16} \qquad \frac{2}{16} - \frac{\sqrt{3}}{16}$$

2.11 Evaluate the integral

$$\int_0^{\pi/12} \sin(3x)\sin x \, dx$$

(A) $\dfrac{3+\sqrt{3}}{12}$ (C) $\dfrac{4-\sqrt{3}}{3}$ (E) $\dfrac{4-\sqrt{2}}{8}$ (G) $\dfrac{4+\sqrt{2}}{8}$

(B) $\dfrac{2+\sqrt{2}}{4}$ (D) $\dfrac{3+\sqrt{2}}{12}$ (F) $\dfrac{2-\sqrt{3}}{16}$ (H) $\dfrac{2+\sqrt{3}}{16}$

• **Use** $\sin(mx)\sin(nx) = \dfrac{1}{2}\big[\cos([m-n]x) - \cos([m+n]x)\big]$

    with $m=3$ $n=1$

$$= \int_0^{\frac{\pi}{12}} \frac{1}{2}\big[\cos 2x - \cos 4x\big]\,dx$$

$$= \frac{1}{2}\left[\frac{1}{2}\sin 2x - \frac{1}{4}\sin 4x\right]_0^{\pi/12}$$

$$= \frac{1}{2}\left[\frac{1}{2}\sin\frac{\pi}{6} - \frac{1}{4}\sin\frac{\pi}{3}\right]$$

$$= \frac{1}{2}\left[\frac{1}{2}\cdot\frac{1}{2} - \frac{1}{4}\cdot\frac{\sqrt{3}}{2}\right]$$

$$= \frac{1}{2}\left[\frac{1}{4} - \frac{\sqrt{3}}{8}\right] = \frac{1}{2}\left[\frac{2-\sqrt{3}}{8}\right]$$

$$= \boxed{\frac{2-\sqrt{3}}{16}}$$

2.12 Evaluate the integral

$$\int_{1/2}^{1} \frac{\sqrt{1-x^2}}{x^2}\, dx$$

(A) 2   (C) 0   (E) $1-\dfrac{\pi}{4}$ (G) $\sqrt{3}-\dfrac{\pi}{3}$

(B) 1   (D) π (F) $\dfrac{\pi}{3}-2$ (H) $\dfrac{\pi}{2}$

$x = 1\,\mathrm{Sin}\ \theta$

$dx = 1\cos\theta\, d\theta$

2.12 Evaluate the integral

$$\int_{1/2}^{1} \frac{\sqrt{1-x^2}}{x^2} \, dx$$

(A) 2   (C) 0   (E) $1-\dfrac{\pi}{4}$ (G) $\sqrt{3}-\dfrac{\pi}{3}$

(B) 1   (D) π (F) $\dfrac{\pi}{3}-2$ (H) $\dfrac{\pi}{2}$

• $\sqrt{a^2-x^2}$: Let $x = a\sin\theta$; $a=1$

Let $x = \sin\theta$
$dx = \cos\theta \, d\theta$
$\sqrt{1-x^2} = \sqrt{1-\sin^2\theta} = \sqrt{\cos^2\theta} = \cos\theta$
$x^2 = \sin^2\theta$

$\int \frac{\sqrt{1-x^2}}{x^2} \, dx = \int \frac{\cos\theta}{\sin^2\theta} \cdot \cos\theta \, d\theta$

$= \int \frac{\cos^2\theta}{\sin^2\theta} \, d\theta = \int \cot^2\theta \, d\theta = \int (\csc^2\theta - 1) \, d\theta$

$= -\cot\theta - \theta$

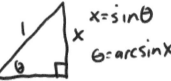

$x = \sin\theta$
$\theta = \arcsin x$

$= \left[-\frac{\sqrt{1-x^2}}{x} - \arcsin x \right]_{1/2}^{1}$ $\frac{\sqrt{\frac{3}{4}}}{\frac{1}{2}} = \frac{\frac{\sqrt{3}}{2}}{\frac{1}{2}} = \sqrt{3}$

$= \left[(0 - \arcsin 1) - \left(-\frac{\sqrt{1-\frac{1}{4}}}{\frac{1}{2}} - \arcsin \frac{1}{2} \right) \right]$

$= -\frac{\pi}{2} + \sqrt{3} + \frac{\pi}{6} = \boxed{\sqrt{3} - \frac{\pi}{3}}$

2.13 Evaluate the integral

$$\int_{0}^{5/2} \frac{12x^2}{\left(25-x^2\right)^{3/2}}\, dx$$

(A) $2\sqrt{3}$ (C) $\dfrac{\pi}{3}$ (E) $\dfrac{\sqrt{3}}{2}+\dfrac{\pi}{4}$ (G) $\sqrt{3}-\dfrac{\pi}{6}$

(B) 1 (D) $\dfrac{\sqrt{3}}{3}$ (F) $4\sqrt{3}-2\pi$ (H) $\dfrac{\pi}{6}$

$$12\int_{0}^{5/2} 25\sin\theta\, d\theta$$

$$12\int_{0}^{5/2} \frac{x^2}{\sqrt{25-x^3}}$$

$$x=$$

$$x = 5\sin\theta$$
$$dx = 5\cos\theta\, d\theta$$
$$x^2 = 25\sin^2\theta$$

$$x^3$$

$$5 \times 5 \times 5$$

$$\sqrt{25 \times 25 \times 25}$$

$$25$$

2.13 Evaluate the integral

$$\int_0^{5/2} \frac{12x^2}{\left(25-x^2\right)^{3/2}}\, dx$$

(A) $2\sqrt{3}$ (C) $\dfrac{\pi}{3}$ (E) $\dfrac{\sqrt{3}}{2}+\dfrac{\pi}{4}$ (G) $\sqrt{3}-\dfrac{\pi}{6}$

(B) 1 (D) $\dfrac{\sqrt{3}}{3}$ (F) $4\sqrt{3}-2\pi$ (H) $\dfrac{\pi}{6}$

• $\sqrt{a^2-x^2}$ Let $x=a\sin\theta$; $a=5$

Let $x=5\sin\theta$ $dx=5\cos\theta\,d\theta$

$(25-x^2)^{3/2}=(25-25\sin^2\theta)^{3/2}=(25(1-\sin^2\theta))^{3/2}$

$=(\sqrt{25\cos^2\theta})^3=(5\cos\theta)^3=5^3\cos^3\theta$

$12x^2=12(25\sin^2\theta)$

$\int\dfrac{12x^2}{(25-x^2)^{3/2}}\,dx=\int\dfrac{12\cdot 25\sin^2\theta}{5^3\cos^3\theta}\,5\cos\theta\,d\theta$

$=12\int\tan^2\theta\,d\theta = 12\int(\sec^2\theta-1)\,d\theta$

$=12(\tan\theta-\theta)$

$=12\left(\dfrac{x}{\sqrt{25-x^2}}-\arcsin\dfrac{x}{5}\right)\Big|_0^{5/2}$

$\sqrt{25-x^2}$

$\sin\theta=\dfrac{x}{5}$

$\theta=\arcsin(x/5)$

$=12\left(\left(\dfrac{5/2}{\sqrt{25-\frac{25}{4}}}-\arcsin\tfrac{1}{2}\right)-0\right)$ $\sqrt{25\cdot\frac{3}{4}}=\dfrac{5\sqrt{3}}{2}$

$=12\left(\dfrac{5/2}{\frac{5}{2}\sqrt{3}}-\dfrac{\pi}{6}\right)=12\left(\dfrac{\sqrt{3}}{3}-\dfrac{\pi}{6}\right)=\boxed{4\sqrt{3}-2\pi}$

2.14 Evaluate the integral

$$\int_0^{1/5} \frac{10}{\left(25x^2+1\right)^2}\, dx$$

(A) $\dfrac{\pi}{4}$ (C) $\dfrac{\pi}{2}$ (E) $\dfrac{1}{4}(\pi+2)$ (G) $\dfrac{1}{2}(\pi+4)$

(B) $\dfrac{\sqrt{2}}{4}$ (D) $\dfrac{\sqrt{2}}{2}$ (F) $\dfrac{1}{2}(\sqrt{2}+1)$ (H) $\dfrac{\sqrt{3}}{2}$

2.14 Evaluate the integral

$$\int_0^{1/5} \frac{10}{\left(25x^2+1\right)^2}\, dx$$

(A) $\dfrac{\pi}{4}$ (C) $\dfrac{\pi}{2}$ (E) $\dfrac{1}{4}(\pi+2)$ (G) $\dfrac{1}{2}(\pi+4)$

(B) $\dfrac{\sqrt{2}}{4}$ (D) $\dfrac{\sqrt{2}}{2}$ (F) $\dfrac{1}{2}(\sqrt{2}+1)$ (H) $\dfrac{\sqrt{3}}{2}$

• $bx^2+a^2 \Rightarrow$ factor out b

$$(25x^2+1)^2 = \left(25\left(x^2+\tfrac{1}{25}\right)\right)^2 = 25^2\left(x^2+\tfrac{1}{25}\right)^2$$

• $x^2+a^2 \Rightarrow$ Let $x=a\tan\theta$; $a=\tfrac{1}{5}$

$x=\tfrac{1}{5}\tan\theta$ $\left(x^2+\tfrac{1}{25}\right)^2 = \left(\tfrac{1}{25}(\tan^2\theta+1)\right)^2$

$dx=\tfrac{1}{5}\sec^2\theta$ $= \left(\tfrac{1}{25}(\sec^2\theta)\right)^2$

$$\int \frac{10}{(25x^2+1)^2}dx = \int \frac{10\cdot\tfrac{1}{5}\sec^2\theta}{25^2\cdot\tfrac{1}{25^2}\cdot\sec^4\theta}\, d\theta$$

$$= 2\int\frac{1}{\sec^2\theta}d\theta = 2\int\cos^2\theta\, d\theta$$

$$= 2\int \tfrac{1}{2}(1+\cos2\theta)\, d\theta = \int(1+\cos2\theta)\, d\theta$$

$$= \theta+\tfrac{1}{2}\sin2\theta = \theta+\tfrac{1}{2}\cdot2\sin\theta\cos\theta$$

$$= \theta+\sin\theta\cos\theta$$

$5x=\tan\theta$

$$=\left[\arctan 5x+\frac{5x}{\sqrt{25x^2+1}}\cdot\frac{1}{\sqrt{25x^2+1}}\right]_0^{1/5}$$

$\sqrt{25x^2+1}$; $5x$; θ ; 1 ; $\theta=\arctan 5x$

$$=\left[\arctan 5x+\frac{5x}{25x^2+1}\right]_0^{1/5}$$

$$= \left(\arctan 1 + \frac{1}{1+1}\right)-(0+0)$$

$$= \frac{\pi}{4}+\frac{1}{2} = \boxed{\frac{1}{4}(\pi+2)}$$

2.15 Solve the differential equation

$$\frac{dy}{dx} = \frac{1}{x^2 \sqrt{x^2 - 9}} \quad \text{with } y(3) = 2.$$

$$dy = \int \frac{1}{x^2 \sqrt{x^2-9}} \, dx$$

$x = 3 \sec\theta$

$dx = 3\sec\theta\tan\theta \, d\theta$

$$\int \frac{1}{9\sec^2\theta \sqrt{9\sec^2\theta - 9}} \cdot 3\sec\theta\tan\theta \, d\theta$$

$\operatorname{arcsec}\left(\frac{x}{3}\right) = \theta$

$$\int \frac{1}{9\sec^2\theta \, 3 \sqrt{\sec^2\theta - 1}}$$

$$\int \frac{1}{9\sec^2\theta \, 3\tan x} \cdot 3s$$

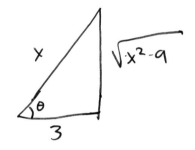

$$\int \frac{3\sec\theta\tan\theta \, d\theta}{9\sec^2\theta \, 3\tan x}$$

$$\frac{1}{9} \int \frac{\sec\theta\tan\theta \, d\theta}{\sec^2\theta} \, d\theta$$

$\sin\theta = \frac{opp}{hyp}$

$\cos\theta = \frac{adj}{hyp}$

$$\frac{1}{9} \int \sec^{-1}\theta \, d\theta$$

$\tan\theta = \frac{opp}{adj}$

$$\frac{1}{9} \int \cos \, d\theta$$

$$\frac{1}{9} \sin\theta + C$$

$$\frac{1}{9} \frac{\sqrt{x^2-9}}{9x} + C \qquad 2 \quad \boxed{C = 2}$$

2.15 Solve the differential equation

$$\frac{dy}{dx} = \frac{1}{x^2\sqrt{x^2-9}} \quad \text{with } y(3) = 2.$$

$$\int dy = \int \frac{1}{x^2\sqrt{x^2-9}} dx$$

$$y = \int \frac{3\sec\theta\tan\theta\, d\theta}{9\sec^2\theta \cdot 3\tan\theta}$$

$$y = \frac{1}{9}\int \frac{1}{\sec\theta}\, d\theta$$

$$y = \frac{1}{9}\int \cos\theta\, d\theta$$

$$y = \frac{1}{9}\sin\theta$$

$$y = \frac{\sqrt{x^2-9}}{9x} + C$$

$$y(3) = 2$$

$$2 = \frac{\sqrt{9-9}}{9\cdot 3} + C \implies C = 2$$

$$\boxed{y = \frac{\sqrt{x^2-9}}{9x} + 2}$$

• $\sqrt{x^2-a^2} \implies$ let $x = a\sec\theta$

$$a = 3$$

$$x = 3\sec\theta$$

$$dx = 3\sec\theta\tan\theta\, d\theta$$

$$x^2 = 9\sec^2\theta$$

$$\sqrt{x^2-9} = \sqrt{9\sec^2\theta - 9}$$
$$= \sqrt{9(\sec^2\theta - 1)}$$
$$= 3\tan\theta$$

$$\frac{x}{3} = \sec\theta = \frac{hyp}{adj}$$

2.16 Find the area of the shaded region below

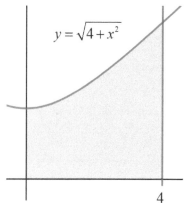

$y = \sqrt{4+x^2}$

4

$$\int_0^4 \sqrt{4+x^2} \; dx$$

$x = 2\tan\theta$

$dx = 2\sec^2\theta \, d\theta$

$$\int_0^4 \sqrt{4 + 4\tan^2\theta}$$

$$2\int_0^4 \sqrt{1+\tan^2\theta}$$

$$2\int_0^4 \sqrt{\sec^2\theta}$$

$$2\int_0^4 \sec x$$

$u = \tan\theta \qquad v = \sec x$

$dx = \sec^2 x \qquad dv = \sec x \tan x$

$$4\int_0^4 \sec^3 x$$

derivative of $\tan x = \sec^2 x$

$\sec x = \sec x \tan x$

$$4\int_0^4 (1+\tan^2\theta)\sec x \, d\theta$$

$$4\int_0^4 u + \tan^2 x \, dx$$

$$4\int_0^4 \sec x + \sec x \tan^2\theta \, d\theta$$

93

2.16 Find the area of the shaded region below

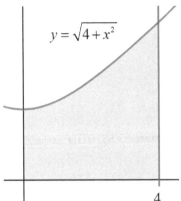

$y = \sqrt{4+x^2}$

4

$A = \int_0^4 \sqrt{4+x^2}\, dx$

$\sqrt{a^2+x^2} \Rightarrow$ let $x = a\tan\theta$

$\sqrt{4+x^2} \Rightarrow$ let $x = 2\tan\theta$

Sub for:

$\boxed{dx = 2\sec^2\theta\, d\theta}$

① dx

② $\sqrt{4+x^2}$

$\sqrt{4+x^2} = \sqrt{4+4\tan^2\theta}$

$= \sqrt{4(1+\tan^2\theta)}$

$= \sqrt{4}\cdot\sqrt{\sec^2\theta}$

$\boxed{\sqrt{4+x^2} = 2\sec\theta}$

$\int_0^4 \sqrt{4+x^2}\, dx = \int 4\sec^3\theta\, d\theta$

$= 4\int \sec^3\theta\, d\theta = 4\cdot\frac{1}{2}\left[\sec\theta\tan\theta + \ln|\sec\theta+\tan\theta|\right]$

Use the reference triangle

$x = 2\tan\theta$

$\tan\theta = \frac{x}{2}\,\frac{opp}{adj}$

$\Rightarrow \sec\theta = \frac{hyp}{adj} = \frac{\sqrt{4+x^2}}{2}$

$\sec\theta\cdot\tan\theta = \frac{x\sqrt{4+x^2}}{4}$

$\sec\theta+\tan\theta = \frac{x+\sqrt{4+x^2}}{2}$

$= 2\left(\frac{x\sqrt{4+x^2}}{4} + \ln\left|\frac{x+\sqrt{4+x^2}}{2}\right|\right)\Big|_0^4$

$= 2\left[\left(\sqrt{20} + \ln\left(\frac{4+\sqrt{20}}{2}\right)\right) - \left(0 + \ln 1\right)\right]_0$

$= 2\left(2\sqrt{5} + \ln\left(2+\frac{2\sqrt{5}}{2}\right)\right)$

$= \boxed{4\sqrt{5} + 2\ln(2+\sqrt{5})}$

Reference box

$\int \sec^3(x)\, dx = \int \sec(x)\sec^2(x)\, dx \qquad u = \sec(x) \qquad dv = \sec^2(x)\, dx$

$du = \sec(x)\tan(x)\, dx \qquad v = \tan(x)$

$\Rightarrow \int \sec(x)\sec^2(x)\, dx = \sec(x)\tan(x) - \int \sec(x)\underbrace{\tan^2(x)}\, dx$

$= \sec(x)\tan(x) - \int \sec(x)\left(\sec^2(x)-1\right)\, dx$

$= \sec(x)\tan(x) - \int \sec^3(x)\, dx + \int \sec(x)\, dx$

$\int \sec^3(x)\, dx = \sec(x)\tan(x) - \int \sec^3(x)\, dx + \int \sec(x)\, dx$

$2\int \sec^3(x)\, dx = \sec(x)\tan(x) + \ln|\sec(x) + \tan(x)|$

$\boxed{\int \sec^3(x)\, dx = \frac{1}{2}\left[\sec(x)\tan(x) + \ln|\sec(x) + \tan(x)|\right] + C}$

2.17 Find the volume of the solid generated by revolving the region bounded by

$$y = \frac{x^2}{\sqrt{9-x^2}}, \ y = 0, \ \text{and } x = 2$$

about the $y-$axis.

Shell

$$2\pi \int x \left(\frac{x^2}{\sqrt{9-x^2}} \right)$$

2.17 Find the volume of the solid generated by revolving the region bounded by

$$y = \frac{x^2}{\sqrt{9-x^2}}, \ y = 0, \text{ and } x = 2$$

about the y axis.

$V_{shell} = 2\pi \int_0^2 x \cdot \frac{x^2}{\sqrt{9-x^2}} dx$

radius $= x$
height $= \frac{x^2}{\sqrt{9-x^2}}$

$V = 2\pi \int_0^2 \frac{x^3}{\sqrt{9-x^2}} dx$

$\sqrt{a^2-x^2}$: Let $x = a\sin\theta$
$x = 3\sin\theta$
$dx = 3\cos\theta d\theta$
$x^3 = 27\sin^3\theta$
$\sqrt{9-x^2} = \sqrt{9(1-\sin^2\theta)}$
$\sqrt{9-x^2} = 3\cos\theta$

$= 2\pi \int \frac{27\sin^3\theta \cdot 3\cos\theta}{3\cos\theta} d\theta$

$= 54\pi \int \sin^3\theta d\theta$

$= 54\pi \int \sin^2\theta \sin\theta d\theta$

$= 54\pi \int (1-\cos^2\theta)\sin\theta d\theta$

$u = \cos\theta$
$du = -\sin\theta d\theta$

$\int (1-u^2)-1 \, du$
$\int (u^2-1) du$
$\frac{u^3}{3} - u$

$= 54\pi \left(\frac{1}{3}(\cos\theta)^3 - \cos\theta \right)$

$= 54\pi \left(\frac{1}{3} \left(\frac{\sqrt{9-x^2}}{3} \right)^3 - \frac{\sqrt{9-x^2}}{3} \right) \Big|^2$

$x = \sin\theta \cdot \frac{1}{3}$

$= 54\pi \left(\left[\frac{1}{3} \cdot \left(\frac{\sqrt{5}}{3} \right)^3 - \frac{\sqrt{5}}{3} \right] - \left[\frac{1}{3} \cdot 1 - 1 \right] \right)^0$

$\cos\theta = \frac{\sqrt{9-x^2}}{3}$

$= 54\pi \left(\frac{\sqrt{5}}{3} \left(\frac{5}{27} - 1 \right) + \frac{2}{3} \right)$

$= 54\pi \left(\frac{2}{3} - \frac{22}{27} \frac{\sqrt{5}}{3} \right)$

$= \frac{54\pi \cdot 2}{3} - \frac{54\pi \, 22\sqrt{5}}{81}$

$= \boxed{36\pi - \frac{44\sqrt{5}}{3}\pi}$

2.18 Evaluate the integral

$$\int_0^1 \frac{5}{2x^2 + 7x + 3} \, dx$$

(A) 1   (C) $\ln\left(\frac{9}{4}\right)$ (E) $\ln 3$ (G) $\ln\left(\frac{3}{2}\right)$

(B) 2   (D) $\ln 2$ (F) $2\ln 3$ (H) The integral diverges

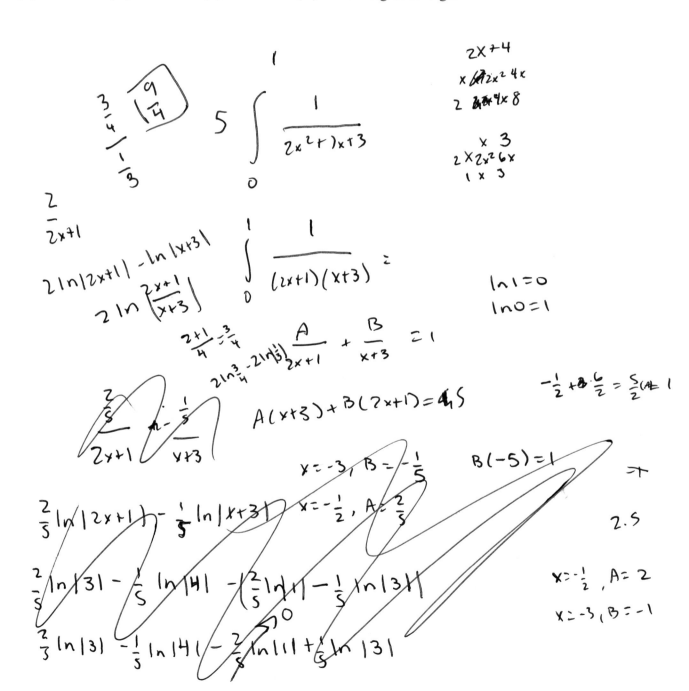

2.18 Evaluate the integral

$$\int_0^1 \frac{5}{2x^2 + 7x + 3} \, dx$$

(A) 1   (C) $\ln\left(\frac{9}{4}\right)$ (E) $\ln 3$ (G) $\ln\left(\frac{3}{2}\right)$

(B) 2   (D) $\ln 2$ (F) $2\ln 3$ (H) The integral diverges

$$2x^2 + 7x + 3 = (2x+1)(x+3)$$

$$\int_0^1 \frac{5 \, dx}{(2x+1)(x+3)} = \int_0^1 \left(\frac{A}{2x+1} + \frac{B}{x+3}\right) dx$$

$$A(x+3) + B(2x+1) = 5$$

let
$x = -3$ $B(-5) = 5$ $B = -1$

let
$x = -\frac{1}{2}$ $A\left(\frac{5}{2}\right) = 5$ $A = 2$

$$= \int_0^1 \left(\frac{2}{2x+1} - \frac{1}{x+3}\right) dx$$

$$= \left[\ln|2x+1| - \ln|x+3|\right]_0^1$$

$$= (\ln 3 - \ln 4) - (\ln 1 - \ln 3)$$

$$= 2\ln 3 - \ln 4$$

$$= \ln 3^2 - \ln 4 = \boxed{\ln\left(\frac{9}{4}\right)}$$

98

2.19 Evaluate the integral

$$\int_0^2 \frac{5x^2 - 2x + 2}{x^{\textcircled{3}}+1}\, dx$$

Hint: $a^3 + b^3 = (a+b)(a^2 - ab + b^2)$

(A) $2\ln 6$ (C) $3\ln 6$ (E) $4\ln 3$ (G) $4\ln 2$

(B) $6\ln 3$ (D) $3\ln 2$ (F) $2\ln 3$ (H) $6\ln 2$

$$\int_0^? \frac{5}{(x+1)} =$$

$$\frac{A}{x} + \frac{B}{1} = 5$$

$$A(1) + B(x) =$$

$$\int_0^? \frac{5x^2 - 2x + 2}{(x+1)(x^2 + x + 1)} \rightarrow$$

$$(x^2 + x + 1)(5 - x + 2)$$

$$\frac{A}{x+1} + \frac{B}{x^2 - x + 1} = 5x^2 - 2x + 2$$

$$A(x^2 - x + 1) + B(x+1) = 5x^2 - 2x + 2$$

$x - 1$

$\frac{x}{H_1}$

① $x_1 = -1$

② $x =$

$$Ax^2 - Ax + A + Bx + B = 5x^2 - 2x + 2$$

x

$$5\ln 1 + 3\ln 1 = \frac{5}{x+1} + \frac{3}{x^2 - x + 1}$$

$A = 5$
$D = 3$
$Bx - 5x = -2x$
$A + B = 2$

$$\left. 5\ln|x+1| + 3\ln|x^2 - x + 1| \right|_0^2$$

$$5\ln 3 + 3\ln 3 = 8\ln 3 - 8\ln 1 \qquad 4 - 2 + 1 = 3$$

$$8\ln\left|\frac{3}{1}\right| = 8\ln 3$$

2.19 Evaluate the integral

$$\int_{0}^{2} \frac{5x^2 - 2x + 2}{x^3 + 1} \, dx$$

Hint: $a^3 + b^3 = (a+b)(a^2 - ab + b^2)$

(A) $2\ln 6$ (C) $3\ln 6$ (E) $4\ln 3$ (G) $4\ln 2$
(B) $6\ln 3$ (D) $3\ln 2$ (F) $2\ln 3$ (H) $6\ln 2$

$$x^3 + 1 = (x+1)(x^2 - x + 1)$$

$$\int_0^2 \frac{5x^2 - 2x + 2}{x^3 + 1} \, dx = \int_0^2 \left(\frac{A}{x+1} + \frac{Bx + C}{x^2 - x + 1} \right) dx$$

$$A(x^2 - x + 1) + (Bx + C)(x + 1) = 5x^2 - 2x + 2$$

let
$x = -1$ $A(1 + 1 + 1) = 5 + 2 + 2$
           $3A = 9$ $A = 3$

let
$x = 0$ $A + C = 2$
           $3 + C = 2$ $C = -1$

let
$x = 1$ $A + 2(B + C) = 5 - 2 + 2$
           $3 + 2(B - 1) = 5$
           $2(B - 1) = 2$
            $B - 1 = 1$
             $B = 2$

$$= \int_0^2 \left(\frac{3}{x+1} + \frac{2x - 1}{x^2 - x + 1} \right) dx \qquad \begin{array}{l} u = x^2 - x + 1 \\ du = (2x - 1)dx \end{array}$$

$$= \left[3\ln|x+1| + \ln|x^2 - x + 1| \right]_0^2$$

$$= (3\ln 3 + \ln 3) - (3\ln 1 + \ln 1)$$

$$= 4\ln 3$$

2.20 Evaluate the integral

$$\int_{\sqrt{3}}^{3} \frac{3x^2 + 3x + 9}{x^3 + 9x} \, dx$$

(A) $\frac{\pi}{3} + \ln\left(\frac{2}{3}\right)$ (C) $\frac{\pi}{6} + \ln\left(\frac{3}{2}\right)$ (E) $\frac{\pi}{12} + \ln\left(\frac{4}{3}\right)$ (G) $\frac{\pi}{15} + \ln 2$

(B) $\frac{\pi}{12} + \ln\left(\frac{3\sqrt{3}}{2}\right)$ (D) $\ln(3) - \frac{\pi}{4}$ (F) $\ln\left(\frac{\sqrt{3}}{2}\right) + \frac{\pi}{6}$ (H) $\frac{\pi}{5} \ln 3$

$$\frac{A}{x} \quad + \quad \frac{Bx + C}{x^2 + 9}$$

$$A(x^2 + 9) + Bx + C(x)$$

$$Ax^2 + 9A + Bx^2 + Cx = 3x^2 + 3x + 9$$

2.20 Evaluate the integral

$$\int_{\sqrt{3}}^{3} \frac{3x^2 + 3x + 9}{x^3 + 9x} \, dx$$

(A) $\frac{\pi}{3} + \ln\left(\frac{2}{3}\right)$ (C) $\frac{\pi}{6} + \ln\left(\frac{3}{2}\right)$ (E) $\frac{\pi}{12} + \ln\left(\frac{4}{3}\right)$ (G) $\frac{\pi}{15} + \ln 2$

(B) $\frac{\pi}{12} + \ln\left(\frac{3\sqrt{3}}{2}\right)$ (D) $\ln(3) - \frac{\pi}{4}$ (F) $\ln\left(\frac{\sqrt{3}}{2}\right) + \frac{\pi}{6}$ (H) $\frac{\pi}{5} \ln 3$

$x^3 + 9x = x(x^2 + 9)$ irreducible quadratic

$\int_{\sqrt{3}}^{3} \frac{3x^2 + 3x + 9}{x^3 + 9x} dx = \int_{\sqrt{3}}^{3} \left(\frac{A}{x} + \frac{Bx + C}{x^2 + 9}\right) dx$

$A(x^2 + 9) + (Bx + C) \cdot x = 3x^2 + 3x + 9$

let $9A = 9$ $A = 1$
$x = 0$

let $10A + (B + C) = 3 + 3 + 9$
$x = 1$ $10 + B + C = 15$
     $B + C = 5$

let $10A + (-B + C)(-1) = 3 - 3 + 9$
$x = -1$ $10 + B - C = 9$
     $B - C = -1$

$\begin{array}{l} B + C = 5 \\ B - C = -1 \\ \hline 2B = 4 \\ B = 2 \end{array}$ $\begin{array}{l} 2 + C = 5 \\ C = 3 \end{array}$

$u = 3x^2$
$dv = 2x \, dx$

$\frac{1}{x^2 + a}$

$\int \frac{1}{u^2 + a}$

$\ln|u^2 + 9|$

$\int_{\sqrt{3}}^{3} \left(\frac{1}{x} + \frac{2x + 3}{x^2 + 9}\right) dx = \int_{\sqrt{3}}^{3} \left(\frac{1}{x} + \frac{2x}{x^2 + 9} + \frac{3}{x^2 + 9}\right) dx$

$= \left[\ln|x| + \ln|x^2 + 9| + 3 \cdot \frac{1}{3} \arctan\left(\frac{x}{3}\right)\right]_{\sqrt{3}}^{3}$

$= (\ln 3 + \ln 18 + \arctan 1) - (\ln\sqrt{3} + \ln 12 + \arctan\frac{\sqrt{3}}{3})$

$= \ln 3 + \ln 18 + \frac{\pi}{4} - \ln\sqrt{3} - \ln 12 - \frac{\pi}{6}$

$= \ln\left(\frac{3 \cdot 18}{\sqrt{3} \cdot 12}\right) + \frac{3\pi - 2\pi}{12} = \ln\left(\frac{9}{2\sqrt{3}}\right) + \frac{\pi}{12}$

$= \ln\left(\frac{9 \cdot \sqrt{3}}{2 \cdot \sqrt{3}\sqrt{3}}\right) + \frac{\pi}{12} = \boxed{\ln\left(\frac{3\sqrt{3}}{2}\right) + \frac{\pi}{12}}$

2.21 Evaluate the integral

$$\int_0^3 \frac{x^3}{x^2 + 6x + 9}\, dx$$

(A) $-27 + 24\ln 2$ (C) $-8 + \ln 2$ (E) $-16 + 24\ln 3$ (G) $-18 + 27\ln 2$

(B) $-12 + 12\ln 3$ (D) $-21 + 24\ln 3$ (F) $-12 + 21\ln 2$ (H) $-24 + 18\ln 3$

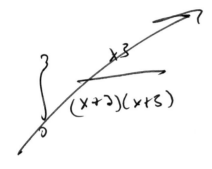

$$\int_0^3 \frac{x^3}{(x+3)(x+3)}$$

$$\begin{array}{r} x - 6 \\ x^2+6x+9 \overline{\smash{\big)}\ x^3 } \\ -\underline{(x^3 + 6x^2 + ax)} \\ -6x^2 + 9x \\ -\underline{(-6x^2 - 36x - 54)} \\ 27x + 27x + 54 \end{array}$$

$$\int_0^3 x + \int_0^3 \frac{27x + 27x + 54}{x^2 + 6x + 9}$$

$$\frac{x^2}{2}$$

$$\frac{9}{2} +$$

2.21 Evaluate the integral

$$\int_0^3 \frac{x^3}{x^2 + 6x + 9}\, dx$$

(A) $-27 + 24\ln 2$ (C) $-8 + \ln 2$ (E) $-16 + 24\ln 3$ (G) $-18 + 27\ln 2$

(B) $-12 + 12\ln 3$ (D) $-21 + 24\ln 3$ (F) $-12 + 21\ln 2$ (H) $-24 + 18\ln 3$

Deg. den < Deg. num ⇒ Long Divide

$$x^2 + 6x + 9 \overline{) x^3 \qquad } \quad \frac{x-6}{}$$

$$-(x^3 + 6x^2 + 9x)$$

$$-6x^2 - 9x$$

$$-(-6x^2 - 36x - 54)$$

$$27x + 54$$

$$= \int_0^3 \left(x - 6 + \frac{27x + 54}{x^2 + 6x + 9} \right) dx$$

$$(x+3)(x+3) = (x+3)^2$$

$$= \int_0^3 \left(x - 6 + \frac{A}{x+3} + \frac{B}{(x+3)^2} \right) dx$$

$$A(x+3) + B = 27x + 54$$

let $x=-3$ $B = -81 + 54 = -27$

let $x=0$ $3A + B = 54$

$$3A - 27 = 54$$

$$3A = 81$$

$$A = 27$$

$$= \int_0^3 \left[x - 6 + 27\left(\frac{1}{x+3} + \frac{-1}{(x+3)^2} \right) \right] dx$$

$$= \left(\frac{x^2}{2} - 6x + 27\left[\ln|x+3| + \frac{1}{x+3} \right] \right) \Big|_0^3$$

$$= \left(\frac{9}{2} - 18 + 27\left[\ln 6 + \frac{1}{6} \right] \right) - \left(0 \cdot 0 + 27\left(\ln 3 + \frac{1}{3} \right) \right)$$

$$= \frac{9}{2} - 18 + 27\ln 6 + \frac{27}{6} - 27\ln 3 - 9$$

$$= 27(\ln 6 - \ln 3) + \frac{9}{2} + \frac{9}{2} - 18 - 9$$

$$= 27\ln\left(\frac{6}{3} \right) + 9 - 18 - 9 = \boxed{27\ln 2 - 18}$$

Section 3: Further Integration Applications

Left Endpoint

$$L_n = \frac{b-a}{n}\left[f\left(x_0\right)+f\left(x_1\right)+\cdots+f\left(x_{n-1}\right)\right]$$

Right Endpoint

$$R_n = \frac{b-a}{n}\left[f\left(x_1\right)+f\left(x_2\right)+\cdots+f\left(x_n\right)\right]$$

Let $\Delta x = \dfrac{b-a}{n}$

$a = x_0$

$b = x_n$

Midpoint

$$M_n = \frac{b-a}{n}\left[f\left(\overline{x_1}\right)+f\left(\overline{x_2}\right)+\cdots+f\left(\overline{x_n}\right)\right]$$

Trapezoid

$$T_n = \frac{\Delta x}{2}\left[f\left(x_0\right)+2f\left(x_1\right)+2f\left(x_2\right)+\cdots+2f\left(x_{n-1}\right)+f\left(x_n\right)\right]$$

Simpson's Rule

$$S_n = \frac{\Delta x}{3}\left(f\left(x_0\right)+4f\left(x_1\right)+2f\left(x_2\right)+4f\left(x_3\right)+\cdots+2f\left(x_{n-2}\right)+4f\left(x_{n-1}\right)+f\left(x_n\right)\right)$$

n must be even for Simpson's Rule

If f'' is continuous and M is an upper bound for $|f''|$ on $[a,b]$, then the error E_T in the trapezoid approximation of the integral from a to b for n subdivisions satifies the inequality

$$\left|E_T\right| \le \frac{M\left(b-a\right)^3}{12n^2}$$

If $f^{(4)}$ is continuous and M is an upper bound for $\left|f^{(4)}\right|$ on $[a,b]$, then the error E_S in the Simpson's approximation of the integral from a to b for n subdivisions satifies the inequality

$$\left|E_S\right| \le \frac{M\left(b-a\right)^5}{180n^4}$$

Improper Integrals

An integral can be called "improper" with one or any combination of the following:

Examples:

- **Infinite upper limit**

$$\int_1^\infty e^{-2x}\,dx = \lim_{t\to\infty}\int_1^t e^{-2x}\,dx$$

- **Infinite lower limit**

$$\int_{-\infty}^1 xe^x\,dx = \lim_{t\to-\infty}\int_t^1 xe^x\,dx$$

- **Infinite discontinuity at:**
  - **upper limit**

$$\int_0^8 \frac{dx}{\sqrt[3]{8-x}} = \lim_{t\to 8^-}\int_0^t \frac{dx}{\sqrt[3]{8-x}}$$

  - **lower limit**

$$\int_0^9 \frac{dx}{\sqrt{x}} = \lim_{t\to 0^+}\int_t^9 \frac{dx}{\sqrt{x}}$$

  - **some value between the upper and lower limit**

$$\int_{-2}^3 \frac{dx}{x^4} = \lim_{t\to 0^-}\int_{-2}^t \frac{dx}{x^4} + \lim_{t\to 0^+}\int_t^3 \frac{dx}{x^4}$$

If the limit exists, we say the integral converges and if it fails to exist (this includes infinite limits), we say the integral diverges.

The skill for evaluating improper integrals relies on the skills of integration and evaluating limits.

Limits at Infinity

If $r > 0$ is a rational number, then $\lim_{x\to\infty}\dfrac{1}{x^r} = 0$

If $r > 0$ is a rational number such that x^r is defined for all x, then $\lim_{x\to-\infty}\dfrac{1}{x^r} = 0$

$f(x)$ is a rational function, with **deg. num. = deg. denom.** $\Rightarrow \lim_{x\to\pm\infty} f(x) = \dfrac{\textbf{coeff. of leading term in num.}}{\textbf{coeff. of leading term in denom.}}$

$f(x)$ is a rational function, with **deg. num. < deg. denom.** $\Rightarrow \lim_{x\to\pm\infty} f(x) = 0$

$f(x)$ is a rational function, with **deg. num. > deg. denom.** $\Rightarrow \lim_{x\to\pm\infty} f(x)$ does not exist

$\left(\text{could be } \infty \text{ or } -\infty\right)$

L'Hopital's Rule

$\lim\limits_{x \to a} \dfrac{f(x)}{g(x)}$ when $f(x) \to 0$ and $g(x) \to 0$ as $x \to a$

$\lim\limits_{x \to a} \dfrac{f(x)}{g(x)} = \dfrac{"0"}{0}$ ← this is called an indeterminate form

$\lim\limits_{x \to a} \dfrac{f(x)}{g(x)}$ when $f(x) \to \pm\infty$ and $g(x) \to \pm\infty$ as $x \to a$

$\lim\limits_{x \to a} \dfrac{f(x)}{g(x)} = \pm \dfrac{"\infty"}{\infty}$ ← this is called an indeterminate form

These two types of indeterminate forms can be simplified using **L'Hopital's Rule**

$$\lim\limits_{x \to a} \dfrac{f(x)}{g(x)} = \lim\limits_{x \to a} \dfrac{f'(x)}{g'(x)}$$ ← assuming that this limit exists

109

Direct Comparison Theorem

Suppose that $f(x)$ and $g(x)$ are continuous functions with $f(x) \geq g(x) \geq 0$ for $x \geq a$.

a) If $\int_a^\infty f(x)\,dx$ is **convergent**, then $\int_a^\infty g(x)\,dx$ is **convergent**.

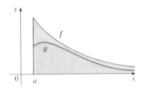

If your function is **smaller** than a function with a **finite** area, then your function will also have finite area

b) If $\int_a^\infty g(x)\,dx$ is **divergent**, then $\int_a^\infty f(x)\,dx$ is **divergent**.

If your function is **larger** than a function with **infinite area**, then your function will also have infinite area.

Two examples worked out on my YouTube Channel:

$$\int_2^\infty \frac{x+1}{\sqrt{x^4 - x}}\,dx$$

$$\int_0^\infty \frac{\arctan x}{2 + e^x}\,dx$$

http://youtu.be/ZYMlAlFeDvc   http://youtu.be/FzXKyf0_b0c

What do you do if the inequality goes the wrong direction?

$$\int_2^\infty \frac{x}{\sqrt{x^4 + 3x}}\,dx$$

$$\frac{x}{\sqrt{x^4 + 3x}} \leq \frac{x}{\sqrt{x^4}} = \frac{1}{x} \quad \text{and} \quad \int_2^\infty \frac{1}{x}\,dx \text{ diverges}$$

So your function is smaller than a function that has infinite area

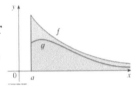

For the direct comparison to work your function needs to be larger than the one with infinite area or smaller than one with finite area.

The good news is that you can still recover for some cases using:

Limit Comparison Theorem

Suppose that $f(x)$ and $g(x)$ are continuous positive functions for $x \geq a$,

and $\lim_{x \to \infty} \frac{f(x)}{g(x)} = L$ with $0 < L < \infty$,

then $\int_a^\infty f(x)\,dx$ and $\int_a^\infty g(x)\,dx$ behave alike.

(They either both converge or they both diverge.)

Probability Density Function

A **random variable**, usually written X, is a variable whose possible values are numerical outcomes of a random phenomenon.

A **continuous random variable** is one which takes an infinite number of possible values (usually measurements)

A **probability density function** is a function f defined for all real x and having the following properties:

1. $f(x) \geq 0$ for all x

2. $\int_{-\infty}^{\infty} f(x)dx = 1$

» Every continuous random variable, X, has a probability density function.

» Used to determine the probability that a continuous random variable lies between two values

$$P(a \leq X \leq b) = \int_{a}^{b} f(x)dx$$

Mean (average value) or expected value of a probability density function $f(x)$ is a measure of the center of a pdf.

$$\mu = \int_{-\infty}^{\infty} xf(x)dx$$

Median (m) of a probability density function $f(x)$ is a number such that ½ the area under the graph of f lies to the right of it (and half the area lies to the left of it).

The median m solves the equation
$$\int_{-\infty}^{m} f(x)dx = \frac{1}{2} \quad \text{or} \quad \int_{m}^{\infty} f(x)dx = \frac{1}{2}$$

Variance $\left(\sigma^2\right)$ of a probability density function is a number that is used to measure the spread of a probability density function $f(x)$.

$$\sigma^2 = \int_{-\infty}^{\infty} (x-\mu)^2 f(x)dx \quad \text{or} \quad \sigma^2 = \int_{-\infty}^{\infty} x^2 f(x)dx - \mu^2$$

Standard Deviation $\left(\sigma\right)$ of a probability density function is a better measurement of spread because the units aren't squared.

$$\sigma = \sqrt{\sigma^2}$$

Separable Differential Equation

$$\frac{dy}{dx} = f(x, y)$$

Where f is a special function that can be factored to be a function of x only multiplied by a function of y only.

$$f(x, y) = g(x) \cdot h(y)$$

1) Take the right hand side and use algebra to represent it as a product of functions one of x only and the other of y only.

$$\frac{dy}{dx} = g(x) \cdot h(y)$$

2) Multiply by dx and divide by $h(y)$

$$\frac{dy}{h(y)} = g(x) dx$$

3) Integrate both sides. Don't forget "+ C" on the right side (technically both sides have it but the two constants can be combined into one.

4) If given an initial condition, immediately plug the x and y in to solve for C.

5) If asked and if possible solve for y in terms of x.

Linear Differential Equation

$$\frac{dy}{dx} + P(x)y = Q(x) \quad \leftarrow \textbf{standard form of a linear first order diff. eq.}$$

1) **Find the integrating factor** $\mu = e^{\int P(x)dx}$

2) **Multiply the entire equation by it.**

$$e^{\int P(x)dx}\frac{dy}{dx} + P(x)e^{\int P(x)dx}y = e^{\int P(x)dx}Q(x)$$

3) **Identify the left hand side as the product rule with the two functions being the integrating factor and the unknown function y(x).**

$$\underbrace{e^{\int P(x)dx}\frac{dy}{dx} + P(x)e^{\int P(x)dx}y}_{\frac{d}{dx}\left(e^{\int P(x)dx} \cdot y\right)} = e^{\int P(x)dx}Q(x)$$

4) **Integrate both sides of the equation.**

$$\int\left[\frac{d}{dx}\left(e^{\int P(x)dx} \cdot y\right)\right] = \int\left[e^{\int P(x)dx}Q(x)\right]dx$$

5) **Use the FTC on the left hand side.**

$$e^{\int P(x)dx} \cdot y = \int\left[e^{\int P(x)dx}Q(x)\right]dx$$

6) **Divide both sides by** μ.

$$y = \frac{\int\left[e^{\int P(x)dx}Q(x)\right]dx}{e^{\int P(x)dx}}$$

⭐ **You can skip steps 2 – 5 and go directly to step 6 from step 1.**

Don't forget the "+ C" in the numerator.

3.1 (a) Estimate the integral below using the Trapezoid rule with 4 trapezoids.

$$\int_{1}^{9} \frac{1}{2} \ln x \, dx$$

(b) Find the number of trapezoids such that the error is at most $\dfrac{1}{100}$.

3.1 (a) Estimate the integral below using the Trapezoid rule with 4 trapezoids.

$$\int_1^9 \frac{1}{2}\ln x\,dx$$

$$\int_a^b f(x)\,dx \approx \frac{\Delta x}{2}\left[f(x_0) + 2f(x_1) + 2f(x_2) + 2f(x_3) + 2f(x_4) + \cdots + 2f(x_{n-1}) + f(x_n)\right]$$

where $\Delta x = \dfrac{b-a}{n}$ and $x_i = a + i\Delta x$.

$$\approx 1 \cdot \left(\ln(315)\right) = \boxed{\ln 315}$$

x	f(x)	multiplier	f(x)*mult.
1	1/2*ln1	1	ln1=0
3	1/2*ln3	2	ln3
5	1/2*ln5	2	ln5
7	1/2*ln7	2	ln7
9	1/2*ln9	1	ln3

$\Delta x = \frac{9-1}{4} = 2$

$\frac{\Delta x}{2} = \frac{2}{2} = 1$

$\frac{1}{2}\ln 9 = \ln(9^{\frac{1}{2}})$
$\quad = \ln 3$

$\overset{+}{\overline{\ln 3 + \ln 5 + \ln 7 + \ln 3}}$

$\ln(3 \cdot 5 \cdot 7 \cdot 3)$

$9 \cdot 7 = 63$

$\times \frac{5}{315}$

(b) Find the number of trapezoids such that the error is at most $\dfrac{1}{100}$.

$|E_T| \le \dfrac{K(b-a)^3}{12n^2} < \dfrac{1}{100}$

n = number of subintervals
b = right endpoint
a = left endpoint
K = maximum value of $|f''(\xi)|$
ξ a number in $[a, b]$ that makes f'' as big as possible

$a = 1$
$b = 9$

$f(x) = \frac{1}{2}\cdot\ln x$
$f'(x) = \frac{1}{2}\cdot\frac{1}{x} = \frac{1}{2}x^{-1}$
$f''(x) = \frac{-1}{2}x^{-2} = \frac{-1}{2x^2}$

$[1,9]\, f''$ max when
$x = 1 \quad f''(1) = -\frac{1}{2}$
$K = |f''(1)| = \frac{1}{2}$

$\dfrac{\frac{1}{2}(9-1)^3}{12n^2} < \dfrac{1}{100}$

$\dfrac{\frac{1}{2}\cdot 8\cdot 8\cdot 8}{12n^2} < \dfrac{1}{100}$

$\dfrac{64}{3n^2} < \dfrac{1}{100}$

$6400 < 3n^2$

$\dfrac{6400}{3} < \dfrac{3n^2}{3}$

$n^2 > 2133.\overline{3}$

$\boxed{n > 47}$

$50^2 = 2500$
$46^2 = 2116$
$47^2 = 2209$

$\begin{array}{cc} 28 & \overset{2}{4}7 \\ 46 & 47 \\ \times 46 & \times 47 \\ \overline{276} & \overline{329} \\ 184 & 188 \\ \overline{2116} & \overline{2209} \end{array}$

3.2 Estimate $\displaystyle\int_{2}^{5} f(x)\,dx$ using the Simpson's rule and 6 subdivisions

for a function $f(x)$ taking on the values in the following table:

x	2	2.5	3	3.5	4	4.5	5
$f(x)$	$\dfrac{1}{2}$	2	1	0	$\dfrac{-3}{2}$	-4	-2

(A) $\dfrac{-5}{12}$ (C) $\dfrac{-11}{8}$ (E) $\dfrac{-7}{12}$ (G) $\dfrac{-13}{12}$

(B) $\dfrac{-11}{12}$ (D) $\dfrac{-5}{8}$ (F) $\dfrac{-13}{8}$ (H) $\dfrac{-7}{4}$

3.2 Estimate $\displaystyle\int_{2}^{5} f(x)\,dx$ using the Simpson's rule and 6 subdivisions

for a function $f(x)$ taking on the values in the following table:

x	2	2.5	3	3.5	4	4.5	5
$f(x)$	$\dfrac{1}{2}$	2	1	0	$\dfrac{-3}{2}$	-4	-2

(A) $\dfrac{-5}{12}$ (C) $\dfrac{-11}{8}$ (E) $\dfrac{-7}{12}$ (G) $\dfrac{-13}{12}$

(B) $\dfrac{-11}{12}$ (D) $\dfrac{-5}{8}$ (F) $\dfrac{-13}{8}$ (H) $\dfrac{-7}{4}$

Simpson's Rule

$$\int_{a}^{b} f(x)\,dx \approx \frac{\Delta x}{3}\Big[f(x_0) + 4f(x_1) + 2f(x_2) + 4f(x_3) + 2f(x_4) + \cdots + 4f(x_{n-1}) + f(x_n)\Big],$$

where $\Delta x = \dfrac{b-a}{n}$ and $x_i = a + i\Delta x$.

	x	f(x)	multiplier	f(x)*mult.
x_0	2	1/2	1	1/2
x_1	5/2	2	4	8
x_2	3	1	2	2
x_3	7/2	0	4	0
x_4	4	-3/2	2	-3
x_5	9/2	-4	4	-16
x_6	5	-2	1	-2

$$\approx \frac{1}{6}\left[\frac{-21}{2}\right] = \frac{-21}{\frac{6\cdot 2}{2}}$$

$$\Delta x = \frac{1}{2}$$

$$\frac{\Delta x}{3} = \frac{\frac{1}{2}}{3} = \frac{1}{6}$$

$$= \boxed{\frac{-7}{4}}$$

$$+$$

$$10\tfrac{1}{2} - 21 = {}^{-}10\tfrac{1}{2} = \frac{-21}{2}$$

3.3 Evaluate the integral

$$\int_{-14}^{2} \frac{dx}{(2-x)^{3/4}}$$

(A) 12   (C) 3   (E) 4   (G) 8
(B) 2   (D) 16   (F) 15   (H) The integral diverges.

3.3 Evaluate the integral

$$\int_{-14}^{2} \frac{dx}{(2-x)^{3/4}}$$

(A) 12   (C) 3   (E) 4   (G) 8

(B) 2   (D) 16   (F) 15   (H) The integral diverges.

Improper integral because $f(2)$ is undef.

$$\lim_{b \to 2} \int_{-14}^{b} \frac{1}{(2-x)^{3/4}} dx$$

$$u = 2-x$$
$$du = -1 \cdot dx$$
$$-1 \cdot du = dx$$
$$-\int u^{-3/4} du$$
$$-4u^{1/4}$$

$$= \lim_{b \to 2} -4(2-x)^{1/4} \Big|_{-14}^{b}$$

$$= \lim_{b \to 2} -4 \left[(2-b)^{1/4} - (16)^{1/4} \right]$$

$$= -4 \left[\underbrace{\lim_{b \to 2}(2-b)^{1/4}}_{\to 0} - 2 \right] = \boxed{8}$$

3.4 Evaluate the integral

$$\int_1^\infty \frac{\ln x}{x^{4/3}}\, dx$$

(A) 4

(B) $\dfrac{4}{9}$

(C) 2

(D) $\dfrac{1}{2}$

(E) $\dfrac{4}{3}$

(F) 9

(G) 1

(H) The integral diverges

3.4 Evaluate the integral

$$\int_1^\infty \frac{\ln x}{x^{4/3}} \, dx$$

(A) 4

(B) $\dfrac{4}{9}$

(C) 2

(D) $\dfrac{1}{2}$

(E) $\dfrac{4}{3}$

(F) 9

(G) 1

(H) The integral diverges

$$\lim_{b\to\infty} \int_1^b \frac{\ln x}{x^{4/3}} \, dx \qquad \text{Integration by Parts}$$

$$uv - \int v\, du \qquad u = \ln x \qquad dv = x^{-4/3}$$
$$du = \frac{1}{x}\, dx \qquad v = -3x^{-1/3}$$

$$= \lim_{b\to\infty} \left. -3x^{-1/3} \ln x \right|_1^b - \int_1^b -3x^{-1/3} \cdot \frac{1}{x}\, dx$$

$$= \lim_{b\to\infty} \left. \frac{-3\ln x}{x^{1/3}} \right|_1^b + 3\int_1^b x^{-4/3}\, dx$$

$$= \lim_{b\to\infty} \frac{-3\ln x}{x^{1/3}} + 3\left[-3x^{-1/3} \right] \Big|_1^b$$

$$= \lim_{b\to\infty} \left(\frac{-3\ln x}{x^{1/3}} - \frac{9}{x^{1/3}} \right) \Big|_1^b$$

$$= \lim_{b\to\infty} \frac{-3\ln b - 9}{b^{1/3}} - \frac{-3\ln 1 - 9}{1} = \boxed{9}$$

$$\underbrace{\phantom{\frac{-3\ln b - 9}{b^{1/3}}}}_{\to 0 \text{ since}}$$

$$\lim_{b\to\infty} \frac{-3\ln b - 9}{b^{1/3}} = \frac{-\infty}{\infty}$$

$$\overset{L'H}{=} \lim_{b\to\infty} \frac{-\frac{3}{b}}{\frac{1}{3} b^{-2/3}} \qquad \frac{-3}{b} \cdot \frac{3b^{2/3}}{1}$$

$$= \lim_{b\to\infty} \frac{-9}{b^{1/3}} = 0$$

3.5 Find the volume of the solid generated by revolving the region in the first quadrant under the curve $y = \dfrac{10}{x^2}$ bounded on the left by

$x = 1$, unbounded on the right, rotated about the $x-$axis.

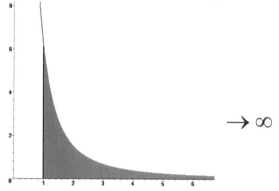

$\rightarrow \infty$

(A) The integral diverges

(B) $\dfrac{25\pi}{2}$

(C) $\dfrac{49\pi}{2}$

(D) $\dfrac{75\pi}{4}$

(E) $\dfrac{10\pi}{3}$

(F) $\dfrac{50\pi}{3}$

(G) $\dfrac{75\pi}{3}$

(H) $\dfrac{100\pi}{3}$

3.5 Find the volume of the solid generated by revolving the region in the first quadrant under the curve $y = \dfrac{10}{x^2}$ bounded on the left by

$x = 1$, unbounded on the right, rotated about the $x-$axis.

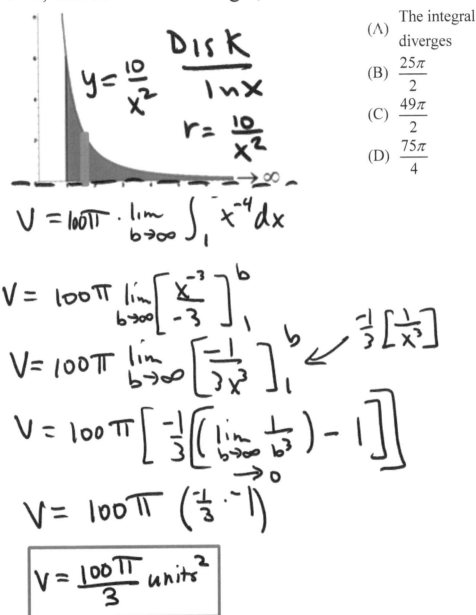

(A) The integral diverges

(B) $\dfrac{25\pi}{2}$

(C) $\dfrac{49\pi}{2}$

(D) $\dfrac{75\pi}{4}$

(E) $\dfrac{10\pi}{3}$

(F) $\dfrac{50\pi}{3}$

(G) $\dfrac{75\pi}{3}$

(H) $\dfrac{100\pi}{3}$

$y = \dfrac{10}{x^2}$

Disk

$\dfrac{}{\ln x}$

$r = \dfrac{10}{x^2}$

$V = 100\pi \cdot \lim_{b \to \infty} \int_1^{\infty} x^{-4}\, dx$

$V = 100\pi \lim_{b \to \infty} \left[\dfrac{x^{-3}}{-3} \right]_1^b$

$V = 100\pi \lim_{b \to \infty} \left[\dfrac{-1}{3x^3} \right]_1^b \quad \leftarrow \quad \dfrac{-1}{3}\left[\dfrac{1}{x^3} \right]$

$V = 100\pi \left[\dfrac{-1}{3}\left(\left(\lim_{b \to \infty} \dfrac{1}{b^3} \right) - 1 \right) \right]$

$\rightarrow 0$

$V = 100\pi \left(\dfrac{-1}{3} \cdot -1 \right)$

$\boxed{V = \dfrac{100\pi}{3} \text{ units}^2}$

3.6 Find the area of the region enclosed by the graphs of
$y = \dfrac{1}{x+1}$ and $y = \dfrac{1}{x+3}$ on the interval $[0,\infty)$.

(A) $\ln 3$

(B) $\ln 4$

(C) 0

(D) 2

(E) $\sqrt{3}$

(F) $\dfrac{1}{2}$

(G) 1

(H) ∞

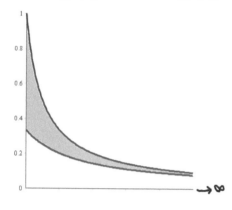

3.6 Find the area of the region enclosed by the graphs of

$y = \dfrac{1}{x+1}$ and $y = \dfrac{1}{x+3}$ on the interval $[0,\infty)$.

(A) $\ln 3$ (E) $\sqrt{3}$

(B) $\ln 4$ (F) $\dfrac{1}{2}$

(C) 0   (G) 1

(D) 2   (H) ∞

$y = \dfrac{1}{x+1}$

$y = \dfrac{1}{x+3}$

@ $x=0$ $y = \dfrac{1}{x+1}$ will be 1 while $y = \dfrac{1}{x+3}$ will be $\dfrac{1}{3}$

$Area = \displaystyle\int_0^\infty \left(\dfrac{1}{x+1} - \dfrac{1}{x+3} \right) dx$

$= \displaystyle\lim_{b\to\infty} \left(\ln|x+1| - \ln|x+3| \right) \Big|_0^b$

$= \displaystyle\lim_{b\to\infty} \left(\ln(b+1) - \ln(b+3) \right) - \left(\underbrace{\ln 1 - \ln 3}_{0} \right)$

$= \displaystyle\lim_{b\to\infty} \ln\left(\dfrac{b+1}{b+3} \right) + \ln 3$

$= \ln\left(\underbrace{\lim_{b\to\infty} \dfrac{b+1}{b+3}}_{\to 1} \right) + \ln 3 = \boxed{\ln 3}$

$\ln 1 = 0$

3.7 Determine whether the integral converges or diverges by using one of the comparison theorems.

I. $\displaystyle\int_{4}^{\infty} \frac{2+\cos x}{\sqrt[3]{x}}\, dx$ II. $\displaystyle\int_{2}^{\infty} \frac{x}{\sqrt{x^5+4}}\, dx$

(A) Both (I) and (II) converge.   (C) (I) converges and (II) diverges.
(B) Both (I) and (II) diverge.   (D) (I) diverges and (II) converges.

3.7 Determine whether the integral converges or diverges by using one of the comparison theorems.

$$\text{I. } \int_{4}^{\infty} \frac{2+\cos x}{\sqrt[3]{x}}\, dx \qquad\qquad \text{II. } \int_{2}^{\infty} \frac{x}{\sqrt{x^5+4}}\, dx$$

(A) Both (I) and (II) converge. (C) (I) converges and (II) diverges.
(B) Both (I) and (II) diverge.   (D) (I) diverges and (II) converges.

I. Compare to $\int_{4}^{\infty} \frac{1}{x^{1/3}}\, dx = \lim_{b\to\infty} \int_{4}^{b} x^{-1/3}\, dx$

Diverges

$= \lim_{b\to\infty} \frac{3}{2}x^{2/3}\Big|_{4}^{b} = \lim_{b\to\infty} \frac{3}{2}\left(b^{2/3} - 4^{2/3}\right) = \infty$

$\frac{1}{x^{1/3}} \le \frac{2+\cos x}{x^{1/3}}$ so $\int_{4}^{\infty} \frac{1}{x^{1/3}}\, dx \le \int_{4}^{\infty} \frac{2+\cos x}{x^{1/3}}\, dx$

∴ By the Direct Comparison Test,

$\int_{4}^{\infty} \frac{2+\cos x}{x^{1/3}}\, dx$ also $\boxed{\text{Diverges}}$

II. Compare to $\int_{2}^{\infty} \frac{x}{x^{5/2}}\, dx = \int_{2}^{\infty} x^{-3/2}\, dx$

$= \lim_{b\to\infty} 2x^{-1/2}\Big|_{2}^{b} = \lim_{b\to\infty} \frac{2}{\sqrt{x}}\Big|_{2}^{b}$

$= \lim_{b\to\infty} \frac{2}{\sqrt{b}} - \frac{2}{\sqrt{2}} = \frac{-2}{\sqrt{2}}$ Converges

$\frac{x}{\sqrt{x^5+4}} \le \frac{x}{\sqrt{x^5}}$ so $\int_{2}^{\infty} \frac{x}{\sqrt{x^5+4}}\, dx \le \int_{2}^{\infty} x^{-3/2}\, dx$

larger denom
⇒ smaller fraction

∴ By the Direct Comparison Test,

$\int_{2}^{\infty} \frac{x}{\sqrt{x^5+4}}\, dx$ also $\boxed{\text{Converges}}$

3.8 Determine whether the integral converges or diverges by using one of the comparison theorems.

I. $\displaystyle\int_{5}^{\infty} \frac{3}{\sqrt{e^x - x}}\, dx$

II. $\displaystyle\int_{1}^{\infty} \frac{\sqrt{x^7 - 5}}{x^4}\, dx$

(A) Both (I) and (II) converge.
(B) Both (I) and (II) diverge.
(C) (I) converges and (II) diverges.
(D) (I) diverges and (II) converges.

3.8 Determine whether the integral converges or diverges by using one of the comparison theorems.

$$\text{I. } \int_{5}^{\infty} \frac{3}{\sqrt{e^x - x}}\, dx \qquad\qquad \text{II. } \int_{1}^{\infty} \frac{\sqrt{x^7 - 5}}{x^4}\, dx$$

(A) Both (I) and (II) converge. (C) (I) converges and (II) diverges.
(B) Both (I) and (II) diverge. (D) (I) diverges and (II) converges.

I. compare to $\int_{5}^{\infty} \frac{3}{\sqrt{e^x}}\, dx = \int_{5}^{\infty} 3 \cdot e^{-x/2}\, dx$

$= \lim_{b \to \infty} 6 e^{-x/2} \Big|_{5}^{b}$

$= \lim_{b \to \infty} 6\left[\frac{1}{e^{b/2}} - \frac{1}{e^{5/2}}\right] = \frac{6}{e^{5/2}}$

Converges

$\frac{3}{\sqrt{e^x}} < \frac{3}{\sqrt{e^x - x}}$ ✓ smaller denom → larger ⟹ wrong Direction for Direct Comp.

Limit Comparison

$\lim_{x \to \infty} \frac{\frac{3}{\sqrt{e^x - x}}}{\frac{3}{\sqrt{e^x}}} = \lim_{x \to \infty} \frac{3}{\sqrt{e^x - x}} \cdot \frac{\sqrt{e^x}}{3} = \lim_{x \to \infty} \sqrt{\frac{e^x}{e^x - x}} = 1$

So they behave alike

∴ By the Limit Comparison Theorem,

$\int_{5}^{\infty} \frac{3}{\sqrt{e^x - x}}\, dx$ also $\boxed{\text{Converges}}$

II. Compare to $\int_{1}^{\infty} \frac{\sqrt{x^7}}{x^4}\, dx = \int_{1}^{\infty} \frac{x^{7/2}}{x^4}\, dx = \int_{1}^{\infty} x^{-1/2}\, dx$

$= \lim_{b \to \infty} 2\sqrt{x}\Big|_{1}^{b} = \lim_{b \to \infty} 2\sqrt{b} - 2 = \infty$

$\frac{\sqrt{x^7 - 5}}{x^4} < \frac{\sqrt{x^7}}{x^4}$ ⟹ wrong Direction for Direct Comp Thm.

smaller numer. & same denom so smaller

Limit Comp. Thm. $\lim_{b \to \infty} \frac{\frac{\sqrt{x^7 - 5}}{x^4}}{\frac{\sqrt{x^7}}{x^4}} = \lim_{b \to \infty} \frac{\sqrt{x^7 - 5}}{x^4} \cdot \frac{x^4}{\sqrt{x^7}}$

$= \lim_{b \to \infty} \frac{\sqrt{x^7 - 5}}{\sqrt{x^7}} = 1$ so they behave alike

∴ By the Limit Comparison Theorem,

$\int_{1}^{\infty} \frac{\sqrt{x^7 + 5}}{x^4}\, dx$ also $\boxed{\text{Diverges}}$

3.9 Consider the probability density function whose graph $f(x)$ is displayed below.

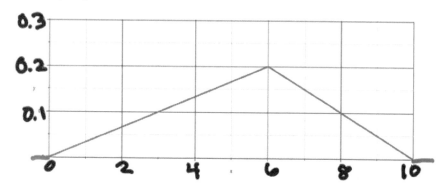

$f(x) = 0$ for $x < 0$ and $x > 10$. Find the probability that $3 \le x \le 8$.

3.9 Consider the probability density function whose graph $f(x)$ is displayed below.

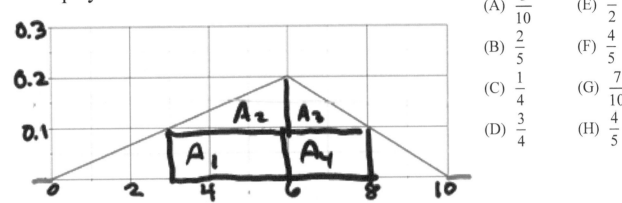

(A) $\dfrac{3}{10}$ (E) $\dfrac{1}{2}$

(B) $\dfrac{2}{5}$ (F) $\dfrac{4}{5}$

(C) $\dfrac{1}{4}$ (G) $\dfrac{7}{10}$

(D) $\dfrac{3}{4}$ (H) $\dfrac{4}{5}$

$f(x)=0$ for $x<0$ and $x>10$. Find the probability that $3 \le x \le 8$.

$$P(3 \le x \le 8) = A_1 + A_2 + A_3 + A_4$$

$$= 0.1(3) + \tfrac{1}{2}(0.1)(3) + \tfrac{1}{2}(0.1)(2) + 0.1(2)$$

$$= 0.3 + \tfrac{1}{2}(0.3) + \tfrac{1}{2}(0.2) + 0.2$$

$$= 0.3 + 0.15 + 0.1 + 0.2$$

$$= \underbrace{0.45} + \underbrace{0.3} = 0.75$$

$$= \boxed{\dfrac{3}{4}}$$

3.10 $f(x)$ below is a probability density function. Find its mean.

$$f(x) = \begin{cases} 16xe^{-4x} & \text{if } x \geq 0 \\ 0 & x < 0 \end{cases}$$

(A) $\dfrac{1}{2}$

(B) $\dfrac{1}{3}$

(C) $\dfrac{1}{4}$

(D) $\dfrac{1}{8}$

(E) $\dfrac{2}{3}$

(F) $\dfrac{3}{4}$

(G) 1

(H) $\dfrac{3}{2}$

3.10 $f(x)$ below is a probability density function. Find its mean.

$$f(x) = \begin{cases} 16xe^{-4x} & \text{if } x \geq 0 \\ 0 & x < 0 \end{cases}$$

(A) $\dfrac{1}{2}$ (E) $\dfrac{2}{3}$

(B) $\dfrac{1}{3}$ (F) $\dfrac{3}{4}$

(C) $\dfrac{1}{4}$ (G) 1

(D) $\dfrac{1}{8}$ (H) $\dfrac{3}{2}$

$M = \int_{-\infty}^{\infty} x \cdot f(x) dx = \int_{0}^{\infty} 16x^2 e^{-4x} dx$

since $f = 0$ for $x < 0$

Use integration by parts shortcut

$$\begin{array}{cc} \underline{D} & \underline{I} \\ 16x^2 & \oplus \quad e^{-4x} \\ 32x & \ominus \quad \frac{1}{4}e^{-4x} \\ 32 & \oplus \quad \frac{1}{16}e^{-4x} \\ 0 & \quad -\frac{1}{64}e^{-4x} \end{array}$$

$= \lim_{b \to \infty} \left[-4x^2 e^{-4x} - 2xe^{-4x} - \frac{1}{2}e^{-4x} \right]_0^b$

$= \lim_{b \to \infty} \left[\dfrac{-4x^2 - 2x - \frac{1}{2}}{e^{4x}} \right]_0^b$

$= \lim_{b \to \infty} \dfrac{-4b^2 - 2b - \frac{1}{2}}{e^{4b}} - \dfrac{-\frac{1}{2}}{1}$

$= \underbrace{\qquad \qquad}_{0} + \quad \frac{1}{2} = \boxed{\dfrac{1}{2}}$

$\lim_{b \to \infty} \dfrac{-4b^2 - 2b - \frac{1}{2}}{e^{4b}} \overset{"-\infty}{=}\overset{}{\infty}" \overset{L'Hop}{=} \lim_{b \to \infty} \dfrac{-8b - 2}{4e^{4b}} \overset{"-\infty}{=}\overset{}{\infty}"$

$\overset{L'Hop}{=} \lim_{b \to \infty} \dfrac{-8}{16e^{4b}} = 0$

3.11 Find the value of k so that the function below is a probability density function.

$$f(x)=\begin{cases} kx^2(3-x) & 0\le x\le 3 \\ 0 & x<0 \text{ or } x>3 \end{cases}$$

(A) $\dfrac{7}{3}$

(B) $\dfrac{1}{3}$

(C) $\dfrac{5}{9}$

(D) $\dfrac{4}{3}$

(E) $\dfrac{2}{3}$

(F) $\dfrac{2}{9}$

(G) $\dfrac{4}{27}$

(H) $\dfrac{11}{27}$

3.11 Find the value of k so that the function below is a probability density function.

$$f(x) = \begin{cases} kx^2(3-x) & 0 \le x \le 3 \\ 0 & x < 0 \text{ or } x > 3 \end{cases}$$

(A) $\dfrac{7}{3}$ (E) $\dfrac{2}{3}$

(B) $\dfrac{1}{3}$ (F) $\dfrac{2}{9}$

(C) $\dfrac{5}{9}$ (G) $\dfrac{4}{27}$

(D) $\dfrac{4}{3}$ (H) $\dfrac{11}{27}$

For a function to be a prob. dens. function,

① $f(x) \ge 0$ for all x

✓ for x<0 or x>3, f=0

for $0 \le x \le 3$

$f = K x^2 (3-x) \Rightarrow K$ must be > 0

② $\displaystyle\int_{-\infty}^{\infty} f(x)\,dx = 1$ since f=0 for x<0 and x>3

$\Rightarrow \displaystyle\int_{0}^{3} Kx^2(3-x)\,dx = 1$ Factor out K and simplify

$K\displaystyle\int_{0}^{3}(3x^2-x^3)\,dx = 1$ Integrate

$K \cdot \left[x^3 - \dfrac{x^4}{4} \right]_{0}^{3} = 1 \Rightarrow K\left[\left(3^3 - \dfrac{3^4}{4}\right) - 0\right] = 1$ Factor out 3^3

$K\left[3^3 \left(1 - \dfrac{3}{4}\right) \right] = 1$

$27K \cdot \dfrac{1}{4} = 1$ mult. by $\dfrac{4}{27}$

$\therefore \boxed{K = \dfrac{4}{27}}$

134

3.12 $f(x)$ below is a probability density function. Find its median.

$$f(x) = \begin{cases} \dfrac{2}{x^3} & \text{if } x \geq 1 \\ 0 & x < 1 \end{cases}$$

(A) 2

(B) $\sqrt{2}$

(C) 4

(D) $\sqrt[3]{3}$

(E) 3

(F) $\sqrt{3}$

(G) 1

(H) $\sqrt[3]{2}$

3.12 $f(x)$ below is a probability density function. Find its median.

$$f(x) = \begin{cases} \dfrac{2}{x^3} & \text{if } x \geq 1 \\ 0 & x < 1 \end{cases}$$

(A) 2   (E) 3

(B) $\sqrt{2}$ (F) $\sqrt{3}$

(C) 4   (G) 1

(D) $\sqrt[3]{3}$ (H) $\sqrt[3]{2}$

Median m is found by solving

$$\int_m^\infty f(x)\,dx = \frac{1}{2} \quad \text{or} \quad \int_{-\infty}^m f(x)\,dx = \frac{1}{2}$$

$f = 0$ for $x < 1$ and $f = \frac{2}{x^3}$ for $x \geq 1$

So this makes the second integral become

$$\int_1^m \frac{2}{x^3}\,dx = \frac{1}{2} \Rightarrow 2\int_1^m x^{-3}\,dx = \frac{1}{2}$$

$$2\left[\frac{x^{-2}}{-2}\right]_1^m = \frac{1}{2} \Rightarrow \left[\frac{-1}{x^2}\right]_1^m = \frac{1}{2}$$

$$\frac{-1}{m^2} - {}^-1 = \frac{1}{2} \Rightarrow \frac{-1}{m^2} + 1 = \frac{1}{2}$$

$$\frac{-1}{m^2} = \frac{-1}{2} \quad \text{or} \quad m^2 = 2 \quad \therefore \boxed{m = \sqrt{2}}$$

136

3.13 Let $y(x)$ be the solution of the differential equation

$$x^3 \frac{dy}{dx} + 2y = e^{1/x^2} \quad \text{with } y(1) = e$$

Find $y\left(\dfrac{1}{2}\right)$.

a) $\dfrac{4}{e^4}$

b) $\dfrac{-1}{2}e^4$

c) $\dfrac{-2}{3}e^4$

d) $\dfrac{1}{e^4}$

e) $\dfrac{1}{2e^4}$

f) $\dfrac{1}{4e^4}$

g) $\dfrac{-3}{2}e^4$

h) $\dfrac{-3}{4}e^4$

137

3.13 Let $y(x)$ be the solution of the differential equation

$$x^3 \frac{dy}{dx} + 2y = e^{1/x^2} \quad \text{with } y(1) = e$$

Find $y\left(\dfrac{1}{2}\right)$.

a) $\dfrac{4}{e^4}$ e) $\dfrac{1}{2e^4}$

b) $\dfrac{-1}{2}e^4$ f) $\dfrac{1}{4e^4}$

c) $\dfrac{-2}{3}e^4$ g) $\dfrac{-3}{2}e^4$

d) $\dfrac{1}{e^4}$ h) $\dfrac{-3}{4}e^4$

This is a linear differential equation.
① Put the equation in Standard form.

$$\frac{dy}{dx} + P(x) \cdot y = Q(x)$$

Divide by x^3 to get

$$\frac{dy}{dx} + \underbrace{\frac{2}{x^3}}_{P(x)} y = \frac{e^{1/x^2}}{x^3}$$

② Find the integrating factor $M = e^{\int P(x)dx}$

$$\int P(x)dx = \int \frac{2}{x^3}dx = \int 2x^{-3}dx = \frac{2x^{-2}}{-2} = \frac{-1}{x^2}$$

$$M = e^{\int P(x)dx} = e^{-\frac{1}{x^2}}$$

③ Mult. the standard form equation by M

$$e^{-\frac{1}{x^2}} \cdot \frac{dy}{dx} + e^{-\frac{1}{x^2}} \cdot \frac{2}{x^3} y = \frac{e^{1/x^2}}{x^3} \cdot e^{-1/x^2}$$

④ Recognize LHS as $\frac{d}{dx}(M \cdot y)$

$$\frac{d}{dx}\left(\underbrace{e^{-\frac{1}{x^2}} \cdot y} \right) = \frac{1}{x^3}$$

⑤ Integrate both sides with respect to x

$$\int \frac{d}{dx}(e^{-\frac{1}{x^2}} \cdot y)dx = \int x^{-3}dx$$

$$e^{-\frac{1}{x^2}} \cdot y = \frac{x^{-2}}{-2} + C$$

$$e^{-\frac{1}{x^2}} \cdot y = \frac{-1}{2x^2} + C$$

⑥ Plug in the initial condition and solve for C $y(1) = e$

$$e^{-\frac{1}{x^2}} \cdot y = \frac{-1}{2x^2} + C$$

$$e^{-1} \cdot e = \frac{-1}{2} + C$$

$$1 = -\frac{1}{2} + C \Rightarrow C = \frac{3}{2}$$

$$e^{-\frac{1}{x^2}} \cdot y = \frac{-1}{2x^2} + \frac{3}{2}$$

⑦ Find $y(\frac{1}{2})$

$$e^{-4} \cdot y = \frac{-1}{2 \cdot \frac{1}{4}} + \frac{3}{2}$$

$$y = \left(-2 + \frac{3}{2}\right) e^4$$

$$\boxed{y = \frac{-1}{2}e^4}$$

3.14 Let $y(x)$ be the solution of the differential equation

$\dfrac{x}{y+2}\cdot\dfrac{dy}{dx}=\dfrac{1}{x+1}$ with $y\left(\dfrac{1}{2}\right)=1$

Find $y(1)$.

a) $\dfrac{1}{2}$

b) $\dfrac{3}{2}$

c) $\dfrac{5}{2}$

d) $\dfrac{7}{2}$

e) $\dfrac{9}{2}$

f) 2

g) 1

h) $\dfrac{3}{4}$

3.14 Let $y(x)$ be the solution of the differential equation

$$\frac{x}{y+2} \cdot \frac{dy}{dx} = \frac{1}{x+1} \quad \text{with } y\left(\frac{1}{2}\right) = 1$$

Find $y(1)$.

a) $\dfrac{1}{2}$ e) $\dfrac{9}{2}$

b) $\dfrac{3}{2}$ f) 2

c) $\dfrac{5}{2}$ g) 1

d) $\dfrac{7}{2}$ h) $\dfrac{3}{4}$

This equation is Separable.

① use algebra to get
y's and | x's and
$\frac{dy}{}$ on LHS | $\frac{dx}{}$ on RHS

$$\frac{1}{y+2} dy = \frac{1}{x+1} \cdot \frac{1}{x} dx$$

② Integrate both sides   use Partial Fractions

$$\int \frac{1}{y+2} dy = \int \frac{1}{x(x+1)} dx$$

$$\int \frac{1}{y+2} dy = \int \left(\frac{A}{x} + \frac{B}{x+1}\right) dx \qquad A(x+1) + Bx = 1$$
$$\text{let } x=0 \quad A=1$$
$$\text{let } x=-1 \quad -B=1$$
$$B=-1$$

$$\int \frac{1}{y+2} dy = \int \left(\frac{1}{x} - \frac{1}{x+1}\right) dx$$

$$\ln|y+2| = \ln|x| - \ln|x+1| + C$$

③ Plug in the initial condition to solve for C
$$y\left(\frac{1}{2}\right) = 1 \Rightarrow x=\frac{1}{2} \text{ & } y=1$$

$$\ln 3 = \ln\left|\frac{1}{2}\right| - \ln\left|\frac{3}{2}\right| + C$$
$$\ln 3 = \ln\left|\frac{1/2}{3/2}\right| + C$$
$$\ln 3 = \ln\left(\frac{1}{3}\right) + C$$
$$C = \ln 3 - \ln\frac{1}{3} \qquad \Rightarrow \underline{C = \ln 9}$$
$$\frac{3}{1/3} = 9$$

④ Find y(1)
$$\ln|y+2| = \ln|x| - \ln|x+1| + \ln 9$$
$$\ln|y+2| = \ln 1 - \ln 2 + \ln 9$$
$$\ln|y+2| = \ln\left(\frac{9}{2}\right)$$
$$\frac{\ln|y+2|}{e} = \frac{\ln(9/2)}{e} \Rightarrow y+2 = \frac{9}{2}$$
$$y = \frac{9}{2} - 2$$
$$\boxed{y = \frac{5}{2}}$$

3.15 Let $y(x)$ be the solution of the differential equation

$$\sqrt{\frac{x}{y}}\,\frac{dy}{dx} = \ln x \quad \text{with } y(1) = 9$$

Find $y(e^2)$.

(A) 2   (C) 3   (E) 4   (G) 8

(B) 10   (D) 16   (F) 25   (H) $2\sqrt{5}$

3.15 Let $y(x)$ be the solution of the differential equation

$$\sqrt{\frac{x}{y}} \frac{dy}{dx} = \ln x \quad \text{with } y(1) = 9$$

Find $y(e^2)$.

(A) 2   (C) 3   (E) 4   (G) 8

(B) 10   (D) 16   (F) 25   (H) $2\sqrt{5}$

The equation is separable.

① use algebra to get
y's and | x's and
dy | dx
on LHS | on RHS

$$\frac{\sqrt{x}}{\sqrt{y}} \frac{dy}{dx} = \ln x$$

$$\frac{1}{\sqrt{y}} dy = \frac{\ln x}{\sqrt{x}} dx$$

② Integrate both sides   ⟵ Integrate by Parts

$$\int y^{-1/2} dy = \int \frac{\ln x}{\sqrt{x}} dx$$

$u = \ln x \quad dv = x^{-1/2}$
$du = \frac{1}{x} dx \quad v = 2\sqrt{x}$

$$2y^{1/2} = 2\sqrt{x} \ln x - 4\sqrt{x} + C$$

$uv - \int v \, du$
$2\sqrt{x} \ln x - 2\int x^{-1/2} dx$

③ Plug in the initial condition to find C $2\sqrt{x} \ln x - 4\sqrt{x}$

$y(1) = 9 \quad x=1 \quad y=9$

$2\sqrt{9} = 2\sqrt{1} \cdot \ln 1 - 4\sqrt{1} + C$

$6 = 2 \cdot 0 - 4 + C$

$\underline{\underline{10 = C}}$

$$2\sqrt{y} = 2\sqrt{x} \ln x - 4\sqrt{x} + 10$$

④ Find $y(e^2)$

$$2\sqrt{y} = 2\sqrt{e^2} \cdot \ln e^2 - 4\sqrt{e^2} + 10$$

$$2\sqrt{y} = 2e \cdot 2 - 4e + 10$$

$$2\sqrt{y} = 4e - 4e + 10$$

$$\sqrt{y} = 5 \Rightarrow \boxed{y = 25}$$

square both sides

3.16 Let $y(x)$ be the solution of the differential equation

$$\left(x^2+1\right)\frac{dy}{dx}+3xy=3x \quad \text{with } y(0)=3$$

Find $y\left(\sqrt{3}\right)$.

(A) $\dfrac{13}{8}$ (C) $\dfrac{3}{2}$ (E) $\dfrac{7}{4}$ (G) $\dfrac{5}{3}$

(B) $\dfrac{1}{2}$ (D) $\dfrac{5}{4}$ (F) $\dfrac{11}{6}$ (H) $\dfrac{11}{8}$

3.16 Let $y(x)$ be the solution of the differential equation

$$(x^2+1)\frac{dy}{dx}+3xy=3x \quad \text{with } y(0)=3$$

Find $y(\sqrt{3})$.

(A) $\dfrac{13}{8}$ (C) $\dfrac{3}{2}$ (E) $\dfrac{7}{4}$ (G) $\dfrac{5}{3}$

(B) $\dfrac{1}{2}$ (D) $\dfrac{5}{4}$ (F) $\dfrac{11}{6}$ (H) $\dfrac{11}{8}$

The equation is both separable and linear.
Let's do this as linear.

$$\frac{dy}{dx} + \underbrace{\frac{3x}{x^2+1}}_{P(x)} \cdot y = \frac{3x}{x^2+1}$$

Find $M = e^{\int P\,dx}$

$$\int \frac{3x}{x^2+1}\,dx \qquad \begin{array}{l} u=x^2+1 \\ du = 2x\,dx \\ \frac{1}{2}du = x\,dx \end{array} \qquad \begin{array}{l} \frac{3}{2}\int \frac{1}{u}\,du \\ = \frac{3}{2}\ln u \\ = \frac{3}{2}\ln(x^2+1) \end{array}$$

$$M = e^{\int P\,dx} = e^{\frac{3}{2}\ln(x^2+1)} = e^{\ln(x^2+1)^{3/2}} = (x^2+1)^{3/2}$$

$$\underbrace{(x^2+1)^{3/2}\frac{dy}{dx} + \frac{3x}{x^2+1}\cdot(x^2+1)^{3/2}\cdot y}_{} = \frac{3x}{x^2+1}(x^2+1)^{3/2}$$

$$\frac{d}{dx}\left((x^2+1)^{3/2}\cdot y\right) = 3x\sqrt{x^2+1} \qquad \begin{array}{l} u=x^2+1 \\ du = 2x\,dx \\ \frac{1}{2}du = x\,dx \\ \frac{3}{2}\int u^{1/2}\,du \\ \cancel{\frac{2}{3}}u^{3/2} \end{array}$$

$$\int \frac{d}{dx}((x^2+1)^{3/2}\cdot y)\,dx = \int 3x\sqrt{x^2+1}\,dx$$

$$(x^2+1)^{3/2}\cdot y = (x^2+1)^{3/2}+C$$

$y(0)=3 \qquad 1\cdot 3 = 1+C \Rightarrow \underline{C=2}$

$$y = 1 + \frac{2}{(x^2+1)^{3/2}}$$

Find $y(\sqrt{3})$

$$y = 1 + \frac{2}{((\sqrt{3})^2+1)^{3/2}} = 1 + \frac{2}{4^{3/2}} = 1+\frac{2}{2^3}$$

$$y = 1 + \frac{1}{2^2} = 1+\frac{1}{4} \Rightarrow \boxed{y=\frac{5}{4}}$$

3.17 Let $y(x)$ be the solution of the differential equation

$$x\frac{dy}{dx} = y + x^2 \sin(x) \quad \text{with } y(\pi) = 0$$

What is $y(2\pi)$?

(a) $-\pi$ (b) -2π (c) -4π (d) 0 (e) 2π (f) 4π

3.17 Let $y(x)$ be the solution of the differential equation

$$x\frac{dy}{dx} = y + x^2 \sin(x) \quad \text{with } y(\pi) = 0$$

What is $y(2\pi)$?

(a) $-\pi$ (b) -2π (c) -4π (d) 0   (e) 2π (f) 4π

Linear

⓪ Put in standard form

$$\frac{dy}{dx} + [P(x)] \, y = Q(x)$$

$$\frac{x\frac{dy}{dx} = y + x^2 \sin x}{x} \quad \begin{array}{l} \cdot \text{divide by } x \\ \cdot \text{subtract over} \\ \quad \text{the } y\text{-term} \\ \cdot \text{put the} - \\ \quad \text{with } P(x) \end{array}$$

$$\underbrace{\frac{dy}{dx} + \left[\frac{-1}{x}\right]}_{P(x)} y = \underbrace{x \sin x}_{Q(x)}$$

① Find $M(x) = e^{\int P(x)dx}$

$$M(x) = e^{\int -\frac{1}{x}dx} = e^{-\ln x}$$

$$M(x) = e^{\ln(x^{-1})} = x^{-1} = \frac{1}{x}$$

Skip to:

$$\boxed{y = \frac{1}{M(x)}\int M(x)Q(x)dx} = x \cdot \int \frac{1}{x} \cdot x \sin x \, dx$$

$$= x \cdot \int \sin x \, dx$$

$$= x \cdot (-\cos x + C)$$

$$y = -x\cos x + Cx \qquad y(\pi) = -\pi\underbrace{\cos \pi}_{-1} + C\pi = 0$$

$$\pi + C\pi = 0 \quad \boxed{C = -1}$$

$$y = -x\cos x - x \qquad y(2\pi) = -(2\pi)\underbrace{\cos(2\pi)}_{1} - 2\pi = \boxed{-4\pi}$$

146

3.18 Let $y(x)$ be the solution of the differential equation

$$xy' + \frac{1}{2}y = x^{3/2} \quad \text{with } y(1) = \frac{5}{2}$$

Find $y(4)$.

(A) 0

(B) e

(C) $\frac{1}{2}$

(D) 1

(E) 2

(F) 3

(G) 4

(H) 5

3.18 Let $y(x)$ be the solution of the differential equation

$xy' + \dfrac{1}{2}y = x^{3/2}$ with $y(1) = \dfrac{5}{2}$

Find $y(4)$.

(A) 0   (E) 2
(B) e (F) 3
(C) $\dfrac{1}{2}$ (G) 4
(D) 1   (H) 5

The equation is linear. Divide by x to put the eq. in std. form

Standard form:

$\dfrac{dy}{dx} + \dfrac{\frac{1}{2}}{x} \cdot y = \dfrac{x^{3/2}}{x}$

$\dfrac{dy}{dx} + \left[\dfrac{1}{2} \cdot \dfrac{1}{x}\right] y = x^{1/2}$

$\underbrace{}_{P(x)}$

$\mu = e^{\int P(x)dx} = e^{\int \frac{1}{2}\cdot\frac{1}{x}dx} = e^{\frac{1}{2}\int \frac{1}{x}dx}$

$\mu = e^{\frac{1}{2}\ln x} = e^{\ln x^{1/2}} = x^{1/2}$

$x^{1/2} \cdot \dfrac{dy}{dx} + \dfrac{1}{2x} \cdot x^{1/2} \, y = x^{1/2} \cdot x^{1/2}$

$\underbrace{}$

$\dfrac{d}{dx}\left[\underset{\mu}{x^{1/2}} \cdot \underset{y}{y}\right] = x$ Now integrate

$\int \dfrac{d}{dx}[x^{1/2} \cdot y] \, dx = \int x \, dx$

$x^{1/2} \cdot y = \dfrac{x^2}{2} + C$

Plug in initial condition to find C
$y(1) = 5/2$ $x=1$ $y = \frac{5}{2}$

$1 \cdot \frac{5}{2} = \frac{1}{2} + C \Rightarrow C = \frac{5}{2} - \frac{1}{2} = 2$

$\sqrt{x} \cdot y = \dfrac{x^2}{2} + 2$

Find y(4).

$\sqrt{4} \cdot y = \dfrac{4^2}{2} + 2 \Rightarrow 2y = \dfrac{8+2}{10}$

$\boxed{y(4) = 5}$

3.19 Let $y(x)$ be the solution of the differential equation

$$\frac{dy}{dx} - xe^{-y} = 2e^{-y} \text{ with } y(0) = 0$$

Find $y(1)$.

(A) $\ln 2$

(B) 1

(C) $\ln\left(\frac{5}{2}\right)$

(D) $\ln\left(\frac{7}{2}\right)$

(E) $\ln\left(\frac{3}{2}\right)$

(F) $\ln 3$

(G) $\ln\left(\frac{9}{2}\right)$

(H) $2\ln 5$

3.19 Let $y(x)$ be the solution of the differential equation

$$\frac{dy}{dx} - xe^{-y} = 2e^{-y} \text{ with } y(0) = 0$$

Find $y(1)$.

(A) $\ln 2$

(C) $\ln\left(\frac{5}{2}\right)$

(E) $\ln\left(\frac{3}{2}\right)$

(G) $\ln\left(\frac{9}{2}\right)$

(B) 1

(D) $\ln\left(\frac{7}{2}\right)$

(F) $\ln 3$

(H) $2\ln 5$

The equation is separable. Solve for $\frac{dy}{dx}$

$$\frac{dy}{dx} = 2e^{-y} + xe^{-y} \quad \text{Factor out } e^{-y}$$

$$\frac{dy}{dx} = e^{-y}(2+x) \Rightarrow \frac{dy}{dx} = \frac{2+x}{e^y}$$

$$e^y dy = (2+x) dx \quad \text{Now integrate}$$

$$\int e^y dy = \int (2+x) dx$$

$$e^y = 2x + \frac{x^2}{2} + C \quad \begin{array}{l} \text{Plug in the} \\ \text{initial cond} \end{array}$$

$$\begin{array}{l} y(0) = 0 \\ e^0 = 0 + 0 + C \Rightarrow C = 1 \end{array}$$

$$e^y = 2x + \frac{x^2}{2} + 1 \quad \text{Find } y(1)$$

$$e^y = 2 + \frac{1}{2} + 1 \Rightarrow e^y = \frac{7}{2}$$

$$\ln(e^y) = \ln\left(\frac{7}{2}\right) \quad \boxed{y(1) = \ln\left(\frac{7}{2}\right)}$$

150

3.20 SET UP (but DO NOT SOLVE) the differential equation for the word problem. Give the differential equation and the initial condition.

A tank initially holds 700 gallons of brine with 12 lbs. of dissolved salt. Brine that contains 5 lbs. of salt per gallon enters the tank at the rate of 4 gallons per minute and the well stirred mixture leaves at the rate of 7 gallons per minute. Let $y(t)$ be amount of salt in the tank at time t. Find the differential equation and initial condition.

3.20 SET UP (but DO NOT SOLVE) the differential equation for the word problem. Give the differential equation and the initial condition.

A tank initially holds 700 gallons of brine with 12 lbs. of dissolved salt. Brine that contains 5 lbs. of salt per gallon enters the tank at the rate of 4 gallons per minute and the well stirred mixture leaves at the rate of 7 gallons per minute. Let $y(t)$ be amount of salt in the tank at time t. Find the differential equation and initial condition.

$$\frac{dy}{dt} = \underbrace{\overbrace{\left[\frac{4 \text{ gal}}{1 \text{ min}}\right]}^{\text{pour rate}} \cdot \overbrace{\left[\frac{5 \text{ lbs.}}{1 \text{ gal}}\right]}^{\text{concentration}}}_{\text{Rate in}} - \underbrace{\overbrace{\left[\frac{7 \text{ gal}}{1 \text{ min}}\right]}^{\text{pour rate}} \cdot \overbrace{\left[\frac{y(t) \text{ lbs.}}{700-3t \text{ gal}}\right]}^{\text{concert.}}}_{\text{Rate out}}$$

Start w/ 700 gal *each min we lose 3 gal.
from the tank
4 gal. enter
7 gal. leave

⟹ Formula for
gallons in tank

$700 - 3t$

$$\boxed{\frac{dy}{dt} = 20 - \frac{7y}{700-3t}}$$

with $y(0) = 12$

initially there are 12 lbs. of dissolved salt

$t = 0$
$y = 12$

3.21 Set up and solve the differential equation for the word problem.

A tank initially holds 100 gallons of brine with 60 lbs. of dissolved salt. Brine that contains 3 lbs. of salt per gallon enters the tank at the rate of 3 gallons per minute and the well stirred mixture leaves at the rate of 3 gallons per minute. Let $y(t)$ be amount of salt in the tank at time t. Find $y(100\ln 2)$.

3.21 Set up and solve the differential equation for the word problem.

A tank initially holds 100 gallons of brine with 60 lbs. of dissolved salt.
Brine that contains 3 lbs. of salt per gallon enters the tank at the rate of 3 gallons per minute and the well stirred mixture leaves at the rate of 3 gallons per minute. Let $y(t)$ be amount of salt in the tank at time t. Find $y(100 \ln 2)$.

$$\frac{dy}{dt} = \underbrace{\overset{\text{Pour rate}}{\left[\frac{3 \text{ gal}}{\text{min}}\right]} \cdot \overset{\text{Conc.}}{\left[\frac{3 \text{ lbs.}}{\text{gal.}}\right]}}_{\text{rate in}} - \underbrace{\overset{\text{Pour rate}}{\left[\frac{3 \text{ gal.}}{\text{min}}\right]} \cdot \overset{\text{Conc.}}{\left[\frac{y \text{ lbs}}{100 \text{ gal.}}\right]}}_{\text{rate out}}$$

no gallons lost or gained

$$\frac{dy}{dt} = 9 - \frac{3y}{100} \quad \longrightarrow \quad \frac{dy}{dt} + \underbrace{\left[\frac{3}{100}\right]}_{P(t)} y = \underbrace{9}_{Q(t)}$$

Linear

$$M = e^{\int P(t)dt} = e^{\frac{3}{100}t}$$

$$y = \frac{1}{M(t)} \int M(t) \cdot Q(t) dt = e^{-\frac{3}{100}t} \int 9 e^{\frac{3}{100}t} dt$$

$$y = e^{-\frac{3}{100}t}\left[9 \cdot \frac{100}{3} e^{\frac{3}{100}t} + C\right] = 300 + Ce^{-\frac{3}{100}t} = y(t)$$

initially $\rightarrow t=0 \quad y = 60 \qquad 60 = 300 + Ce^0 \Rightarrow C = -240$

$$\boxed{y = 300 - 240 e^{-\frac{3}{100}t}} \qquad y(100 \ln 2) = 300 - 240 e^{-\frac{3}{100} \cdot 100 \ln 2}$$

$$y(100 \ln 2) = 300 - 240 e^{-3\ln 2} = 300 - 240 e^{\ln(2^{-3})}$$

$$= 300 - \frac{240}{8} = 300 - 30 = \boxed{270 \text{ lbs.}}$$

3.22 Use Newton's Law of Cooling to set up and solve the differential equation for the word problem.

The brewing pot temperature of coffee is 200°F and the room temperature is 72°F. After 5 minutes the temperature of the coffee is 180°F. How long ill it take for the coffee to reach a serving temperature of 150°F?

3.22 Use Newton's Law of Cooling to set up and solve the differential equation for the word problem.

The brewing pot temperature of coffee is 200°F and the room temperature is 72°F. After 5 minutes the temperature of the coffee is 180°F. How long ill it take for the coffee to reach a serving temperature of 150°F?

Newton's Law of Cooling: The rate at which an object's temperature is changing at any given time is roughly proportional to the difference between its temperature and the temperature of the surrounding medium

y = temp. (in °F)   Surrounding medium = Room temperature = R
t = time (in min.) K = constant of proportionality

$$\frac{dy}{dt} = K(R-y)$$

$R = 72$ $\frac{dy}{dt} = K(72-y)$ Separable

$t = 0$
$y = 200$ $\int \frac{dy}{72-y} = \int K\,dt$

$t = 5$
$y = 180$ $-1 \cdot \ln|72-y| = Kt + C$

$$\ln|72-y| = -Kt - C$$

$$e^{\ln|72-y|} = e^{-Kt} \cdot e^{-C} \qquad e^{-C} = A$$

$$72-y = e^{-Kt} \cdot e^{-C}$$

$$72-y = Ae^{-Kt} \Rightarrow \boxed{y = 72 - Ae^{-Kt}}$$

$y(0) = 200$ $200 = 72 - A \cdot e^0 \Rightarrow A = -128$

$y = 72 + 128e^{-Kt}$ $y(5) = 180$

$180 = 72 + 128e^{-5K}$ $\frac{180-72}{128} = e^{-5K}$

$\frac{27}{32} = e^{-5K}$ $\ln\left(\frac{27}{32}\right) = -5K \Rightarrow K = -\frac{1}{5}\ln\left(\frac{27}{32}\right)$

$$\boxed{y(t) = 72 + 128 e^{\frac{1}{5}\ln\left(\frac{27}{32}\right)\cdot t}} \qquad y = 150 \quad t = ?$$

$150 = 72 + 128 e^{\frac{1}{5}\ln\left(\frac{27}{32}\right)t}$

$\frac{78}{128} = e^{\frac{1}{5}\ln\left(\frac{27}{32}\right)t} \rightarrow \ln\left(\frac{39}{64}\right) = \frac{1}{5}\ln\left(\frac{27}{32}\right)t$

$t = \dfrac{\ln\left(\frac{39}{64}\right)}{\frac{1}{5}\ln\left(\frac{27}{32}\right)} \approx \boxed{14.6 \text{ min.}}$

Section 4: Sequences and Series

Finding the limit of a sequence boils down
to being able to find limits at infinity.

Tools :

Limit Laws   **Limits at Infinity**

Indeterminate forms and L'Hopitals Rule

Thoerems :

1. Squeeze Theorem:

$$\left.\begin{array}{c} a_n \le c_n \le b_n \text{ for all } n > N \\ and \\ \lim_{n \to \infty} a_n = \lim_{n \to \infty} b_n = L \end{array}\right\} \Rightarrow \lim_{n \to \infty} c_n = L$$

2. $\lim_{n \to \infty} |a_n| = 0 \Rightarrow \lim_{n \to \infty} a_n = 0$

3. The sequence $\{r^n\}$ is $\begin{cases} \text{convergent to 0} & \text{if } -1 < r < 1 \\ \text{convergent to 1} & r = 1 \\ \text{divergent} & \text{for all other values of } r \end{cases}$

4. $\left.\begin{array}{c} \lim_{n \to \infty} a_n = L \\ and \\ f \text{ is contin. at } L \end{array}\right\} \Rightarrow \lim_{n \to \infty} f(a_n) = f\left(\lim_{n \to \infty} a_n\right) = f(L)$

   bring the limit inside

5. Every **bounded** and **increasing** sequence and every **bounded** and **decreasing** sequence is **convergent**.

div.   **conv. to 0**   div.

A **geometric series** is one in which each term is obtained from the preceding one by multiplying it by the common ratio r.

$$a + ar + ar^2 + ar^3 + \cdots + ar^{n-1} + \cdots = \sum_{n=1}^{\infty} ar^{n-1}$$

this only converges for certain values of r.

The geometric series $\sum_{n=1}^{\infty} ar^{n-1}$ converges to t

The sum $s = \lim_{n \to \infty} S_n = \dfrac{a}{1-r} = \dfrac{\text{first term}}{1 - \text{ratio}}$ provided that $-1 < r < 1$ or $|r| < 1$.

The geometric series diverges for all other values of r

div.   **conv. to** $\dfrac{a}{1-r}$ div.

A **telescoping series** is one in which the middle terms cancel and the sum collapses into just a few terms.

Example:

$$\sum_{n=1}^{\infty}\left(\frac{3}{n^2}-\frac{3}{(n+1)^2}\right)$$

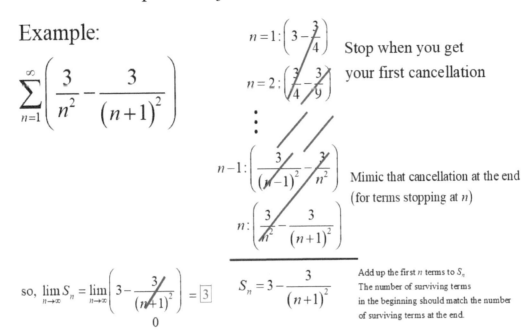

Stop when you get your first cancellation

Mimic that cancellation at the end (for terms stopping at n)

$$S_n = 3 - \frac{3}{(n+1)^2}$$

Add up the first n terms to S_n
The number of surviving terms in the beginning should match the number of surviving terms at the end.

so, $\lim_{n\to\infty} S_n = \lim_{n\to\infty}\left(3 - \frac{3}{(n+1)^2}\right) = \boxed{3}$

If the series $\sum_{n=1}^{\infty} a_n$ is convergent, then $\lim_{n\to\infty} a_n = 0$.

Converse:

If $\lim_{n\to\infty} a_n = 0$, then the series $\sum_{n=1}^{\infty} a_n$ is convergent. This is **False!**

$\left(\text{Just because } \lim_{n\to\infty} a_n = 0, \text{ you } \textbf{cannot} \text{ conclude that the series } \sum_{n=1}^{\infty} a_n \text{ is convergent.}\right)$

$$\boxed{\lim_{n\to\infty}\frac{1}{n} = 0, \text{ but } \sum_{n=1}^{\infty}\frac{1}{n} \text{ diverges.}}$$

Contrapositive:

Test for Divergence:

If $\lim_{n\to\infty} a_n \neq 0$ or $\lim_{n\to\infty} a_n$ does not exist, then the series $\sum_{n=1}^{\infty} a_n$ is divergent.

This is **True!**

The Integral Test

If $f(x)$ is: $a)$ continuous,   on the interval $[k, \infty)$
constant $k>0$
$b)$ positive,
$c)$ and decreasing

, then the series $\sum\limits_{n=k}^{\infty} a_n$ $\left(\text{with } a_n = f(n)\right)$

$i)$ is convergent when $\int\limits_{k}^{\infty} f(x)\,dx$ is convergent.

$ii)$ is divergent when $\int\limits_{k}^{\infty} f(x)\,dx$ is divergent.

Note:

the function does not necessarily have to be decreasing for all $x \in [k, \infty)$
as long as the function is decreasing "eventually"
$\left(\text{there is some number } N \text{ so that } f \text{ is decreasing for all } x > N\right)$

$f(x) = \dfrac{1}{x^p}$

on $[1, \infty)$

$a)$ continuous,
$b)$ positive,
$c)$ and decreasing

For what values of p does the integral converge?

$$\int\limits_{1}^{\infty} \frac{1}{x^p}\,dx = \lim_{b \to \infty} \int\limits_{1}^{b} x^{-p}\,dx = \lim_{b \to \infty} \left.\frac{x^{-p+1}}{-p+1}\right|_{1}^{b}$$

need $-p+1$ to be negative so that
we can get convergence by moving
the x – term to the denominator

$$-p+1 < 0 \Rightarrow \boxed{p > 1}$$

corresponding to this function is the series $\sum\limits_{n=1}^{\infty} \dfrac{1}{n^p}$

this is called a $\underline{p-series}$

$i)$ $\sum\limits_{n=1}^{\infty} \dfrac{1}{n^p}$ **Converges** when $p > 1$ $ii)$ $\sum\limits_{n=1}^{\infty} \dfrac{1}{n^p}$ **Diverges** when $p \leq 1$

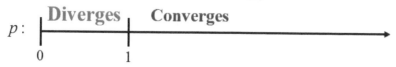

161

Remainder Estimate for the Integral Test

If $f(x)$ is: $a)$ continuous, on the interval $[k, \infty)$
$\qquad\qquad$ $b)$ positive, $\qquad\qquad$ constant $k>0$
$\qquad\qquad$ $c)$ and decreasing $\qquad\qquad\left(\text{with } a_n = f(n)\right)$

and the series $\displaystyle\sum_{n=k}^{\infty} a_n$ is convergent.

, then the remainder $R_n = s_n - s$ can be bounded above by $\displaystyle\int_{n}^{\infty} f(x)dx$.

$$s = \underbrace{a_1 + a_2 + a_3 + \cdots + a_n}_{s_n} + \underbrace{a_{n+1} + a_{n+2} + a_{n+3} + \cdots}_{R_n} \qquad \left(\text{and below by } \int_{n+1}^{\infty} f(x)dx.\right)$$

$$\Rightarrow R_n = a_{n+1} + a_{n+2} + a_{n+3} + \cdots$$

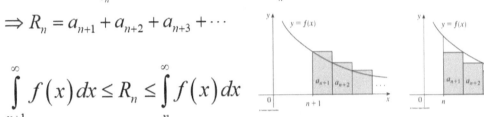

$$\int_{n+1}^{\infty} f(x)dx \le R_n \le \int_{n}^{\infty} f(x)dx$$

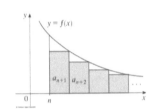

The Comparison Tests

<u>The Direct Comparison Test:</u>

Given the series $\sum\limits_{n=1}^{\infty} a_n$, $(a_n \geq 0)$

   (i) if the terms a_n are smaller than the terms b_n of a known **convergent**

series $\sum\limits_{n=1}^{\infty} b_n$ $(b_n \geq 0)$, then our series $\sum\limits_{n=1}^{\infty} a_n$ is also **convergent**.

   (ii) if the terms a_n are larger than the terms b_n of a known **divergent**

series $\sum\limits_{n=1}^{\infty} b_n (b_n \geq 0)$, then our series $\sum\limits_{n=1}^{\infty} a_n$ is also **divergent**.

For the series $\sum\limits_{n=1}^{\infty} b_n$, it must be known whether it converges or diverges, so

it is usually chosen to be a <u>$p-$series</u> or a <u>geometric series</u>.

      search for the **dominating terms** in both the numerator and the denominator of a_n,

      choose your b_n to be the ratio of these dominating terms

<u>The Limit Comparison Test:</u>

Given the series $\sum\limits_{n=1}^{\infty} a_n$, $(a_n > 0)$ and a known

convergent or divergent series $\sum\limits_{n=1}^{\infty} b_n$, $(b_n > 0)$

  1. If the $\lim\limits_{n \to \infty} \dfrac{a_n}{b_n} = c$ where c is a finite positive number, then

    the series will behave alike, i.e. either both converge or both diverge.

  2. If the $\lim\limits_{n \to \infty} \dfrac{a_n}{b_n} = 0$ **and** $\sum b_n$ converges, then

    the series will behave alike, i.e. $\sum a_n$ also converges.

  3. If the $\lim\limits_{n \to \infty} \dfrac{a_n}{b_n} = \infty$ **and** $\sum b_n$ diverges, then

    the series will behave alike, i.e. $\sum a_n$ also diverges.

The Ratio Test

Let $\{a_n\}$ be a sequence and assume that the following limit exists: $\lim\limits_{n\to\infty}\left|\dfrac{a_{n+1}}{a_n}\right| = L$

i) If $L < 1$, then the series $\sum\limits_{n=1}^{\infty} a_n$ is absolutely convergent.

ii) If $L > 1$ or if the limit is infinite, then the series $\sum\limits_{n=1}^{\infty} a_n$ is divergent.

iii) If $L = 1$, the Ratio Test is inconclusive.

(the series could be absolutely convergent, conditionally convergent, or divergent)

The Root Test

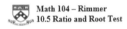

Let $\{a_n\}$ be a sequence and assume that the following limit exists: $\lim\limits_{n\to\infty}\sqrt[n]{|a_n|} = L$

i) If $L < 1$, then the series $\sum\limits_{n=1}^{\infty} a_n$ is absolutely convergent.

ii) If $L > 1$ or if the limit is infinite, then the series $\sum\limits_{n=1}^{\infty} a_n$ is divergent.

iii) If $L = 1$, the Root Test is inconclusive.

(the series could be absolutely convergent, conditionally convergent, or divergent)

The Alternating Series Test

If the alternating series $\sum_{n=1}^{\infty}(-1)^{n-1}b_n$ $\left(\text{where } b_n > 0\right)$ satisfies:

$i)$ $\lim_{n\to\infty} b_n = 0$

$ii)$ $\{b_n\}$ is a decreasing sequence, and

,then the series is **convergent**.

Note:

$a)$ This test is for convergence only. It says nothing about divergence.

$b)$ Like the function in the Integral Test, the sequence $\{b_n\}$ needs to be decreasing "eventually" i.e., for all $n > N$ for some N

Alternating Series Estimation Theorem

If the alternating series $\sum_{n=1}^{\infty}(-1)^{n-1}b_n$ $\left(\text{where } b_n > 0\right)$ satisfies:

$i)$ $\lim_{n\to\infty} b_n = 0$

$ii)$ $\{b_n\}$ is a decreasing sequence

The sum $s = S_n + R_n$

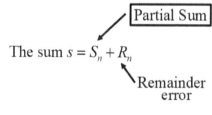

then $|R_n| = |s - S_n| \le b_{n+1}$

The size of the error is at most the size of the first omitted term.
$$b_{n+1}$$

The actual sum is between $S_n - b_{n+1}$ and $S_n + b_{n+1}$.
$$|s - S_n| \le b_{n+1} \Rightarrow -b_{n+1} \le s - S_n \le b_{n+1}$$
$$S_n - b_{n+1} \le s \le S_n + b_{n+1}$$

The error has the same sign as the first omitted term.
$$(-1)^n b_{n+1} > 0 \Rightarrow R_n > 0 \Rightarrow s - S_n > 0 \quad \Rightarrow s > S_n \Rightarrow S_n \text{ is a underestimate}$$
$$(-1)^n b_{n+1} < 0 \Rightarrow R_n < 0 \Rightarrow s - S_n < 0 \quad \Rightarrow s < S_n \Rightarrow S_n \text{ is a overestimate}$$

Absolute Convergence

An infinite series

$$\sum_{n=1}^{\infty} a_n \text{ is called \textbf{absolutely convergent} if the positive series } \sum_{n=1}^{\infty} |a_n| \text{ converges.}$$

Absolute convergence implies converges.
(If the series of absolute value converges, then the original series also converges)

If the series of absolute value **diverges**, it is still possible
for the original series to converge.

Use the Alternating Series Test on the original series.
If the Alternating Series Test gives convergence, then this is a special
type of convergence.

An infinite series

$$\sum_{n=1}^{\infty} a_n \text{ is called \textbf{conditionally convergent} if it converges but } \sum_{n=1}^{\infty} |a_n| \text{ diverges.}$$

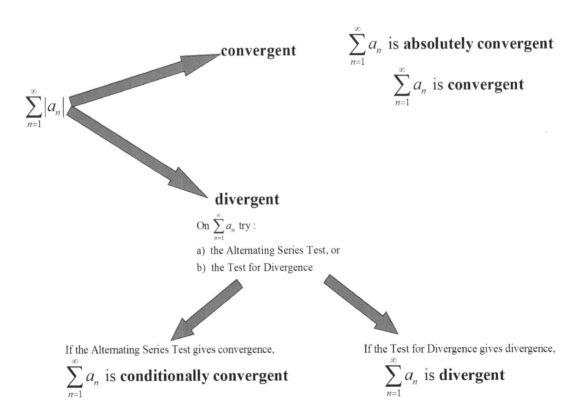

convergent

$$\sum_{n=1}^{\infty} a_n \text{ is \textbf{absolutely convergent}}$$

$$\sum_{n=1}^{\infty} a_n \text{ is \textbf{convergent}}$$

$$\sum_{n=1}^{\infty} |a_n|$$

divergent

On $\sum_{n=1}^{\infty} a_n$ try :

a) the Alternating Series Test, or
b) the Test for Divergence

If the Alternating Series Test gives convergence,

$$\sum_{n=1}^{\infty} a_n \text{ is \textbf{conditionally convergent}}$$

If the Test for Divergence gives divergence,

$$\sum_{n=1}^{\infty} a_n \text{ is \textbf{divergent}}$$

A series of the form

$$\sum_{n=0}^{\infty} c_n(x-a)^n = c_0 + c_1(x-a) + c_2(x-a)^2 + \ldots$$

is called a **power series** centered at a or a power series about a

We use the Ratio Test (or the Root Test) to find for what values of x the series converges.

$$\lim_{n \to \infty} \left| \frac{a_{n+1}}{a_n} \right| = L < 1 \text{ for convergence}$$

solve for $|x-a|$ to get $|x-a| < R$

R is called the **radius of convergence** (R.O.C.).

$$\Rightarrow -R < x-a < R$$

$$\Rightarrow a-R < x < a+R$$

use square brackets [or]

This is called the **interval of convergence** (I.O.C.).

Plug in the endpoints to check for convergence or divergence at the endpoints.

use parentheses (or)

$$f(x) = \sum_{n=0}^{\infty} \frac{f^{(n)}(a)}{n!}(x-a)^n = f(a) + f'(a)(x-a) + \frac{f''(a)}{2!}(x-a)^2 + \frac{f'''(a)}{3!}(x-a)^3 + \cdots$$

Taylor series of the function f centered at a.

If $a = 0$, then we call the series the **Maclaurin series** of the function f.

$$f(x) = \sum_{n=0}^{\infty} \frac{f^{(n)}(0)}{n!}x^n = f(0) + f'(0)x + \frac{f''(0)}{2!}x^2 + \frac{f'''(0)}{3!}x^3 + \cdots$$

List of important Maclaurin series :

$$\frac{1}{1-x} = \sum_{n=0}^{\infty} x^n = 1 + x + x^2 + x^3 + \cdots \quad IOC = (-1,1)$$

$$\frac{1}{(1-x)^2} = \sum_{n=1}^{\infty} nx^{n-1} = 1 + 2x + 3x^2 + 4x^3 + \cdots \quad IOC = (-1,1)$$

$$\ln(1-x) = -\sum_{n=0}^{\infty} \frac{x^{n+1}}{n+1} = -x - \frac{x^2}{2} - \frac{x^3}{3} - \frac{x^4}{4} - \cdots \quad IOC = (-1,1)$$

$$\arctan x = \sum_{n=0}^{\infty} (-1)^n \frac{x^{2n+1}}{2n+1} = x - \frac{x^3}{3} + \frac{x^5}{5} - \frac{x^7}{7} + \cdots \quad IOC = (-1,1)$$

$$e^x = \sum_{n=0}^{\infty} \frac{x^n}{n!} = 1 + x + \frac{x^2}{2!} + \frac{x^3}{3!} + \frac{x^4}{4!} + \cdots \quad IOC = (-\infty, \infty)$$

$$\sin x = \sum_{n=0}^{\infty} \frac{(-1)^n x^{2n+1}}{(2n+1)!} = x - \frac{x^3}{3!} + \frac{x^5}{5!} - \frac{x^7}{7!} + \cdots \quad IOC = (-\infty, \infty)$$

$$\cos x = \sum_{n=0}^{\infty} \frac{(-1)^n x^{2n}}{(2n)!} = 1 - \frac{x^2}{2!} + \frac{x^4}{4!} - \frac{x^6}{6!} + \cdots \quad IOC = (-\infty, \infty)$$

Taylor Remainder Estimate

Given a function $f(x)$ and an interval $[b,c]$ with $a \in (b,c)$

(a is the center of the Taylor series for $f(x)$)

If there is a positive constant M such that $\left| f^{(n+1)}(t) \right| \le M$,

(M is the upper bound on the $(n+1)$st derivative of $f(x)$)

then the Remainder in Taylor Series satifies

$$|R_n| \le M \frac{|x-a|^{n+1}}{(n+1)!}$$

n is the term that you stopped at for your approximation.

Choose a t in $[b,c]$ that will cause the $(n+1)$st derivative to be as big as possible.

Choose an x in $[b,c]$ that will cause the $|x-a|^{n+1}$ to be as big as possible.

4.1 Determine the limit of each sequence

a) $a_n = \left\{ \sqrt{n^2 + 3n} - n \right\}$

b) $a_n = \left\{ \dfrac{e^n}{1 + e^{-n}} \right\}$

4.1 Determine the limit of each sequence

a) $a_n = \left\{\sqrt{n^2 + 3n} - n\right\}$ $a_n = \left\{\dfrac{e^n}{1 + e^{-n}}\right\}$

$\sqrt{n^2+3n+n}$

Multiply by the conjugate

$$\lim_{n\to\infty} \frac{\sqrt{n^2+3n}-n}{1} \cdot \frac{\sqrt{n^2+3n}+n}{\sqrt{n^2+3n}+n} = \lim_{n\to\infty} \frac{n^2+3n-n^2}{\sqrt{n^2+3n}+n}$$

$$= \lim_{n\to\infty} \frac{3n}{\sqrt{n^2+3n}+n} \cdot \frac{\frac{1}{n}}{\frac{1}{n}} = \lim_{n\to\infty} \frac{3}{\dfrac{\sqrt{n^2+3n}}{n} + 1}$$

$\uparrow$
rename to
$\sqrt{n^2}$

$$= \lim_{n\to\infty} \frac{3}{\sqrt{\dfrac{n^2+3n}{n^2}} + 1}$$

$$= \lim_{n\to\infty} \frac{3}{\sqrt{1 + \underbrace{\frac{3}{n}}_{\to 0 \text{ as } n\to\infty}} + 1} = \frac{3}{\sqrt{1}+1} = \boxed{\dfrac{3}{2}}$$

b) $a_n = \left\{\dfrac{e^n}{1 + e^{-n}}\right\}$

$$\lim_{n\to\infty} \frac{e^n}{1+e^{-n}} = \lim_{n\to\infty} \frac{e^n}{1 + \underbrace{\frac{1}{e^n}}_{\to 0 \text{ as } n\to\infty}} = \boxed{\infty}$$

170

4.2 Determine the limit of the sequence

$$a_n = \left\{ n \left[\ln(n+3) - \ln(n) \right] \right\}$$

(a) 0   (b) 1   (c) $\ln(3)$ (d) 3   (e) ∞ (f) the limit does not exist

$$n \ln\left(\frac{n+3}{n}\right)$$

$$n \ln \frac{1+\frac{3}{n}}{1}$$

$$\frac{-\frac{1}{n}}{-\frac{1}{n^2}}$$

$$n \ln\left(\frac{n+3}{n}\right)$$

4.2 Determine the limit of the sequence

$$a_n = \{n[\ln(n+3) - \ln(n)]\}$$

(a) 0   (b) 1   (c) $\ln(3)$ (d) 3   (e) ∞ (f) the limit does not exist

$$\lim_{n \to \infty} n[\ln(n+3) - \ln(n)] = \lim_{n \to \infty} n \cdot \ln\left(\frac{n+3}{n}\right)$$

$$= \lim_{n \to \infty} n \cdot \ln\left(1 + \frac{3}{n}\right) = ``\infty \cdot 0" \quad \text{as } n \to \infty, \frac{3}{n} \to 0$$

$$= \lim_{n \to \infty} \frac{\ln\left(1 + \frac{3}{n}\right)}{\frac{1}{n}} = \frac{``0"}{0}$$

Instead of multiplying by n, divide by 1/n. This creates a fraction that should be L'Hopital ready

$$\stackrel{L'H}{=} \lim_{n \to \infty} \frac{\frac{1}{1 + \frac{3}{n}} \cdot 3 \cdot \frac{-1}{n^2}}{\frac{-1}{n^2}} = \lim_{n \to \infty} \frac{3}{1 + \frac{3}{n}}$$

$$= \boxed{3}$$

4.3 Determine the limit of the sequence

$$a_n = \left(\cos\left(\tfrac{1}{n}\right)\right)^{n^2}$$

a) e

c) $\dfrac{-1}{2}$

e) $\dfrac{1}{\sqrt{e}}$

g) $\sqrt{e}$

b) e^2

d) 1

f) 0

h) ∞

4.3 Determine the limit of the sequence

$$a_n = \left(\cos\left(\tfrac{1}{n}\right)\right)^{n^2}$$

a) e c) $\dfrac{-1}{2}$ e) $\dfrac{1}{\sqrt{e}}$ g) $\sqrt{e}$

b) e^2 d) 1 f) 0 h) ∞

$$\lim_{n\to\infty}\left(\cos\left(\tfrac{1}{n}\right)\right)^{n^2} = "1^\infty"$$

$$y = \lim_{n\to\infty}\left(\cos\left(\tfrac{1}{n}\right)\right)^{n^2}$$

$$\ln y = \lim_{n\to\infty}\ln\left(\cos\left(\tfrac{1}{n}\right)\right)^{n^2}$$

$$\ln y = \lim_{n\to\infty} n^2\ln\left(\cos\left(\tfrac{1}{n}\right)\right) = "\infty\cdot 0"$$

Instead of multiplying by n^2, divide by 1/n^2. This creates a fraction that should be L'Hopital ready

$$\ln y = \lim_{n\to\infty}\frac{\ln\left(\cos\left(\tfrac{1}{n}\right)\right)}{\frac{1}{n^2}} = "\frac{0}{0}" \quad \text{use L'Hop.}$$

$$\ln y = \lim_{n\to\infty}\frac{\frac{1}{\cos\left(\tfrac{1}{n}\right)}\cdot -\sin\left(\tfrac{1}{n}\right)\cdot \frac{-1}{n^2}}{\frac{-2}{n^3}} = \lim_{n\to\infty}\frac{-\tan\left(\tfrac{1}{n}\right)\cdot\frac{-1}{n^2}}{\frac{2}{n}\cdot\frac{-1}{n^2}}$$

$$\ln y = \lim_{n\to\infty}\frac{-\tan\left(\tfrac{1}{n}\right)}{\frac{2}{n}} = "\frac{0}{0}" \quad \text{use L'Hop. again}$$

$$\ln y = \lim_{n\to\infty}\frac{-\sec^2\left(\tfrac{1}{n}\right)\cdot\frac{-1}{n^2}}{2\cdot\frac{-1}{n^2}} = \lim_{\substack{n\to\infty \\ \frac{1}{n}\to 0}}\frac{-\sec^2\left(\tfrac{1}{n}\right)}{2} = \frac{-1}{2}$$

$$\ln y = -\tfrac{1}{2} \qquad e^{\ln y} = e^{-\tfrac{1}{2}} \Rightarrow \boxed{y = e^{-\tfrac{1}{2}}}$$

174

4.4 Determine the limit of the sequence

$$a_n = \left\{ \left(2 + \frac{n^2}{2} \right)^{\frac{1}{\ln(2n)}} \right\}$$

A) ∞ B) $\dfrac{1}{e}$ C) e^2 D) 0

E) 1 F) e G) $\dfrac{1}{e^2}$ H) $\sqrt{e}$

4.4 Determine the limit of the sequence

$$a_n = \left\{ \left(2 + \frac{n^2}{2} \right)^{\frac{1}{\ln(2n)}} \right\}$$

A) ∞ B) $\dfrac{1}{e}$ C) e^2 D) 0

E) 1 F) e G) $\dfrac{1}{e^2}$ H) $\sqrt{e}$

$$\lim_{n\to\infty} \left(2 + \frac{n^2}{2} \right)^{\frac{1}{\ln 2n}} = \infty^{0} \quad \text{indeterminant power}$$

$$y = \lim_{n\to\infty} \left(2 + \frac{n^2}{2} \right)^{\frac{1}{\ln(2n)}} \quad \text{take } \ln$$

$$\ln y = \lim_{n\to\infty} \ln \left[\left(2 + \frac{n^2}{2} \right)^{\frac{1}{\ln(2n)}} \right]$$

$$\ln y = \lim_{n\to\infty} \frac{1}{\ln(2n)} \cdot \ln \left(2 + \frac{n^2}{2} \right) = 0 \cdot \infty$$

$$\ln y = \lim_{n\to\infty} \frac{\ln \left(2 + \frac{n^2}{2} \right)}{\ln(2n)} = \frac{\infty}{\infty} \quad \text{use L'Hôpital's Rule}$$

$$\ln y \overset{L'H}{=} \lim_{n\to\infty} \frac{\frac{1}{2 + \frac{n^2}{2}} \cdot n}{\frac{1}{2n} \cdot 2} = \lim_{n\to\infty} \frac{\frac{n}{2 + \frac{n^2}{2}}}{\frac{1}{n}}$$

$$\ln y = \lim_{n\to\infty} \frac{n}{2 + \frac{n^2}{2}} \cdot \frac{n}{1} = \lim_{n\to\infty} \frac{n^2}{2 + \frac{n^2}{2}} \cdot \frac{\frac{1}{n^2}}{\frac{1}{n^2}} = \lim_{n\to\infty} \frac{1}{\frac{2}{n^2} + \frac{1}{2}} = 2$$

$$\underset{\to 0}{}$$

$$\ln y = 2 \qquad e^{\ln y} = e^2 \Rightarrow y = e^2$$

$$\boxed{\lim_{n\to\infty} \left(2 + \frac{n^2}{2} \right)^{\frac{1}{\ln(2n)}} = e^2}$$

4.5 Determine the limit of the sequence

$$a_n = \left\{ \left(\frac{n+2}{n-2} \right)^n \right\}$$

a) e c) e^4 e) 2   g) 0

b) e^2 d) 1   f) 4   h) ∞

4.5 Determine the limit of the sequence

$$a_n = \left\{ \left(\frac{n+2}{n-2} \right)^n \right\}$$

a) e c) e^4 e) 2   g) 0

b) e^2 d) 1   f) 4   h) ∞

$$\lim_{n \to \infty} \left(\frac{n+2}{n-2} \right)^n = "1^\infty" \text{ Indeterminant Power}$$

$$y = \lim_{n \to \infty} \left(\frac{n+2}{n-2} \right)^n \quad \text{Take } \ln \text{ of both sides}$$

$$\ln y = \lim_{n \to \infty} \ln \left(\frac{n+2}{n-2} \right)^n$$

$$\ln y = \lim_{n \to \infty} n \cdot \ln \left(\frac{n+2}{n-2} \right) = "\infty \cdot 0"$$

$$\ln y = \lim_{n \to \infty} \frac{\ln \left(\frac{n+2}{n-2} \right)}{\frac{1}{n}} = "\frac{0}{0}"$$

Before using L'Hôpital's Rule, simplify

$$\ln y = \lim_{n \to \infty} \frac{\ln(n+2) - \ln(n-2)}{\frac{1}{n}}$$

$$\ln y \overset{L'H}{=} \lim_{n \to \infty} \frac{\frac{1}{n+2} - \frac{1}{n-2}}{\frac{-1}{n^2}} = \lim_{n \to \infty} \frac{\frac{n-2-(n+2)}{n^2-4}}{\frac{-1}{n^2}}$$

$$\ln y = \lim_{n \to \infty} \frac{-4}{n^2-4} \cdot \frac{-n^2}{1} = \lim_{n \to \infty} \frac{4n^2}{n^2-4} = 4$$

deg num = deg den so the limit is the ratio of coeff on highest deg. terms

$$\ln y = 4$$

$$e^{\ln y} = e^4 \implies y = e^4$$

$$\boxed{\lim_{n \to \infty} \left(\frac{n+2}{n-2} \right)^n = e^4}$$

4.6 Determine whether the series converges or diverges. If it converges, find its sum.

$$\sum_{n=4}^{\infty} \frac{4}{n^2-1}$$

$\frac{1}{n^2}$

$4 \sum_{n=4}^{\infty} \frac{1}{n^2-1} \mathscr{S}$

$\frac{4}{(n+1)^2-1} \cdot \frac{n^2-1}{4}$

$\frac{4^1}{2n}$

$\begin{matrix} & & n+1 \\ n & n^2 & n \\ 1 & n & \times \end{matrix}$

$\frac{4}{6} =$

$\frac{4}{15} + \cdots \frac{4}{24} + \frac{4}{35} + \frac{4}{48}$

$\dfrac{\frac{4}{15}}{1-\frac{1}{}}$

4.6 Determine whether the series converges or diverges. If it converges, find its sum.

$$\sum_{n=4}^{\infty} \frac{4}{n^2 - 1}$$

$n^2 - 1 = (n+1)(n-1)$ use partial fractions

$$\sum_{n=4}^{\infty} \frac{4}{n^2-1} = \sum_{n=4}^{\infty} \left(\frac{A}{n+1} + \frac{B}{n-1} \right) \quad A(n-1) + B(n+1) = 4$$

$\quad n=1: \quad 2B = 4 \Rightarrow B = 2$

$\quad n=-1 \quad -2A = 4 \Rightarrow A = 2$

$$= \sum_{n=4}^{\infty} \left(\frac{2}{n-1} - \frac{2}{n+1} \right) \quad \text{Telescoping Series}$$

$$\lim_{n\to\infty} S_n = \lim_{n\to\infty} \underbrace{\left(\frac{2}{3} - \frac{2}{5} \right)}_{n=4} + \underbrace{\left(\frac{2}{4} - \frac{2}{6} \right)}_{n=5} + \underbrace{\left(\frac{2}{5} - \frac{2}{7} \right)}_{n=6} + \cdots$$

$$\cdots + \underbrace{\left(\frac{2}{n-3} - \frac{2}{n-1} \right)}_{n-2} + \underbrace{\left(\frac{2}{n-2} - \frac{2}{n} \right)}_{n-1} + \underbrace{\left(\frac{2}{n-1} - \frac{2}{n+1} \right)}_{n}$$

♦ $\frac{2}{5}$ was the first to cancel. Higher terms $\frac{2}{6}$ and $\frac{2}{7}$ will cancel

Mimic the first cancellation with the end terms

$\frac{2}{n-1}$ cancels. This is the final term to cancel.

Terms lower than it will cancel

$$\lim_{n\to\infty} S_n = \lim_{n\to\infty} \frac{2}{3} + \frac{2}{4} - \underbrace{\frac{2}{n} - \frac{2}{n+1}}_{\substack{\to 0 \text{ as} \\ n\to\infty}} = \frac{2}{3} + \frac{1}{2}$$

$$= \frac{4+3}{6}$$

$$\boxed{Sum = \frac{7}{6}}$$

4.7 Determine whether the series converges or diverges.

If it converges, find its sum.

$$\sum_{n=1}^{\infty}\left[\left(\frac{6}{7}\right)^{n}-\frac{3}{2^{n}}\right]$$

A) converges, sum is 5   B) converges, sum is 3   C) converges, sum is 2
D) converges, sum is 6   E) converges, sum is 4   F) converges, sum is 7
G) diverges   H) none of the above

$$\frac{6^{n}}{7^{n}}-\frac{3}{2^{n}}$$

4.7 Determine whether the series converges or diverges.

If it converges, find its sum.

$$\sum_{n=1}^{\infty}\left[\left(\frac{6}{7}\right)^n - \frac{3}{2^n}\right]$$

A) converges, sum is 5   B) converges, sum is 3   C) converges, sum is 2
D) converges, sum is 6   E) converges, sum is 4   F) converges, sum is 7
G) diverges   H) none of the above

$$= \sum_{n=1}^{\infty}\left(\frac{6}{7}\right)^n - \sum_{n=1}^{\infty}\frac{3}{2^n}$$ Two Geometric Series

Sum1   Sum2

★ Sum1 $= \frac{6}{7} + \frac{36}{49} + \cdots$

first term $= \frac{6}{7}$

ratio $= \frac{6}{7}$ or r $|r|<1$ so the sum conv. to:

$$Sum1 = \frac{\frac{6}{7}}{1-\frac{6}{7}} = \frac{6/7}{1/7} = \frac{6}{7}\cdot\frac{7}{1} = \boxed{6}$$

★ Sum2 $= 3\cdot\sum_{n=1}^{\infty}\frac{1}{2^n} = 3\cdot\sum_{n=1}^{\infty}\left(\frac{1}{2}\right)^n$

first term $= 3\cdot\frac{1}{2}$

ratio $= \frac{1}{2}$ or r $|r|<1$ so the sum conv. to:

$$Sum2 = \frac{3/2}{1-1/2} = \frac{3}{2}\cdot\frac{2}{1} = \boxed{3}$$
                          $\underbrace{}_{1/2}$

Answer: Sum1 $-$ Sum2   (since they both conv.)

$$6 - 3 = \boxed{3}$$

4.8 Find the sum of the series

$$\sum_{n=1}^{\infty} \frac{1}{n(n+2)}$$

A) 3 / 4 $_2 (^{4})$ B) 1 / 2   C) 3 / 5   D) 9 / 10
E) 2 / 3   F) 4 / 5   G) divergent   H) none of the above

$$\frac{1}{3} + \frac{1}{8} + \frac{1}{15} \qquad \left\{ \frac{1}{n} \left\{ \frac{1}{n+2} \right. \right.$$

4.8 Find the sum of the series

$$\sum_{n=1}^{\infty} \frac{1}{n(n+2)}$$

A) $3/4$ B) $1/2$ C) $3/5$ D) $9/10$

E) $2/3$ F) $4/5$ G) divergent   H) none of the above

This is a telescoping series

$$= \sum_{n=1}^{\infty} \left(\frac{A}{n} + \frac{B}{n+2} \right) \qquad A(n+2) + Bn = 1$$

$$n=0 \quad 2A = 1 \Rightarrow A = \tfrac{1}{2}$$
$$n=-2 \quad -2B = 1 \Rightarrow B = -\tfrac{1}{2}$$

$$= \sum_{n=1}^{\infty} \left(\frac{1/2}{n} - \frac{1/2}{n+2} \right)$$

$$\text{Sum} = \lim_{n \to \infty} S_n = \lim_{n \to \infty} \underbrace{\left(\frac{1/2}{1} - \frac{1/2}{3} \right)}_{n=1} + \underbrace{\left(\frac{1/2}{2} - \frac{1/2}{4} \right)}_{n=2} + \underbrace{\left(\frac{1/2}{3} - \frac{1/2}{5} \right)}_{} + \cdots$$

$$+ \cdots \underbrace{\left(\frac{1/2}{n-2} - \frac{1/2}{n} \right)}_{n-2} + \underbrace{\left(\frac{1/2}{n-1} - \frac{1/2}{n+1} \right)}_{n-1} + \underbrace{\left(\frac{1/2}{n} - \frac{1/2}{n+2} \right)}_{n}$$

↠ $\frac{1/2}{3}$ was the first to cancel. Higher terms $\frac{1/2}{4}$ and $\frac{1/2}{5}$ will cancel

Mimic the first cancellation with the end terms

$\frac{1/2}{n}$ cancels. This is the final term to cancel.

Terms lower than it will cancel.

$$\text{Sum} = \lim_{n \to \infty} \frac{1}{2} + \frac{1}{4} - \underbrace{\frac{1/2}{n+1} - \frac{1/2}{n+2}}_{\to 0 \text{ as } n \to \infty} = \frac{1}{2} + \frac{1}{4}$$

$$\boxed{\text{Sum} = \frac{3}{4}}$$

4.9 Determine whether the series converges or diverges.
If it converges, find its sum.

$$\sum_{n=1}^{\infty} \left[(0.6)^{n-1} - (0.2)^{n} \right]$$

4.9 Determine whether the series converges or diverges.

If it converges, find its sum.

$$\sum_{n=1}^{\infty} \left[(0.6)^{n-1} - (0.2)^n \right]$$

We have two geometric series

$$= \sum_{n=1}^{\infty} (0.6)^{n-1} - \sum_{n=1}^{\infty} (0.2)^n$$

$$\underbrace{\qquad}_{sum\,1} \qquad \underbrace{\qquad}_{sum\,2}$$

★ sum 1 :

first term = 1

ratio = 0.6 ∈ r |r|<1 so the

series conv. to :

$$Sum\,1 = \frac{1}{1-0.6} = \frac{1}{0.4} = \frac{1}{\frac{4}{10}} = \frac{10}{4} = \boxed{\frac{5}{2}}$$

★ sum 2 :

first term = 0.2

ratio = 0.2 ∈ r |r|<1 so the

series conv. to :

$$Sum\,2 = \frac{0.2}{1-0.2} = \frac{0.2}{0.8} = \frac{2}{8} = \boxed{\frac{1}{4}}$$

The original series will conv. to :

$$Sum\,1 - Sum\,2 = \frac{5}{2} - \frac{1}{4} = \frac{10-1}{4}$$

$$\boxed{Sum = \frac{9}{4}}$$

4.10 Determine whether the series converges or diverges.
If it converges, find its sum.

$$\frac{1}{e^2} - \frac{2\pi}{e^5} + \frac{4\pi^2}{e^8} - \frac{8\pi^3}{e^{11}} + \cdots$$

1   2   3   4

$$\frac{(2\pi)^{n-1}}{e^{(3n-1)}}$$

$$r = \frac{-2\pi}{e^3}$$

$$\frac{-2\pi}{e^3}$$

$$\frac{\frac{1}{e^2}}{1 - \left(-\frac{2\pi}{e^3}\right)}$$

4.10 Determine whether the series converges or diverges.

If it converges, find its sum.

$$\frac{1}{e^2} - \frac{2\pi}{e^5} + \frac{4\pi^2}{e^8} - \frac{8\pi^3}{e^{11}} + \cdots$$

$\times \frac{-2\pi}{e^3} \quad \times \frac{-2\pi}{e^3} \quad \times \frac{-2\pi}{e^3}$ Geometric Series

$r = \frac{-2\pi}{e^3} \qquad |r| = \frac{2\pi}{e^3} < 1$ since

$\underline{\text{Conv. to}} \qquad\qquad 2\pi < e^3$

$\qquad\qquad\qquad\qquad \sim 6 \qquad \sim 9$

$\text{Sum} = \dfrac{\frac{1}{e^2}}{1 - \left(\frac{-2\pi}{e^3}\right)} = \dfrac{\frac{1}{e^2}}{1 + \frac{2\pi}{e^3}} \cdot \dfrac{e^3}{e^3} = \dfrac{e}{e^3 + 2\pi}$

$$\boxed{\text{Sum} = \dfrac{e}{e^3 + 2\pi}}$$

4.11 Determine whether the series converges or diverges.
If it converges, find its sum.

$$\sum_{n=1}^{\infty}\left(\arccos\left(\frac{\sqrt{3}}{n+1} \right) - \arccos\left(\frac{\sqrt{3}}{n+2} \right) \right)$$

$$\arccos \frac{\dfrac{\sqrt{3}}{n+1}}{\dfrac{\sqrt{3}}{n+2}}$$

$$\frac{\sqrt{3}(n+2)}{\sqrt{3}(n+1)}$$

4.11 Determine whether the series converges or diverges.
If it converges, find its sum.

$$\sum_{n=1}^{\infty}\left(\arccos\left(\frac{\sqrt{3}}{n+1}\right)-\arccos\left(\frac{\sqrt{3}}{n+2}\right)\right)$$

This is a telescoping series.

$Sum = \lim_{n\to\infty} S_n = \left(\arccos\frac{\sqrt{3}}{2} - \arccos\frac{\sqrt{3}}{3}\right) + \left(\arccos\frac{\sqrt{3}}{3} - \arccos\frac{\sqrt{3}}{4}\right) + \dots$

$\qquad n=1 \qquad\qquad\qquad n=2$

$\qquad \dots + \left(\arccos\frac{\sqrt{3}}{n} - \arccos\frac{\sqrt{3}}{n+1}\right) + \left(\arccos\frac{\sqrt{3}}{n+1} - \arccos\frac{\sqrt{3}}{n+2}\right)$

$\qquad\qquad n-1 \qquad\qquad\qquad n$

★ $\arccos\frac{\sqrt{3}}{3}$ was the first to cancel. Higher term $\arccos\frac{\sqrt{3}}{4}$ will cancel

Mimic the first cancellation with the end terms
$\arccos\frac{\sqrt{3}}{n+1}$ cancels. This is the final term to cancel.
Terms lower than it will cancel.

$Sum = \lim_{n\to\infty} \underbrace{\arccos\frac{\sqrt{3}}{2}} - \arccos\frac{\sqrt{3}}{n+2}$

$\qquad\qquad\qquad\qquad\qquad\qquad \to 0 \text{ as } n\to\infty$

$\arccos\frac{\sqrt{3}}{\infty}$, $\frac{\sqrt{3}}{\infty} \to \cos^{-1}0$, $\arccos 0 = 0$

$Sum = \frac{\pi}{6} - \frac{\pi}{2} = \frac{\pi - 3\pi}{6} = \frac{-2\pi}{6}$

$$\boxed{Sum = -\frac{\pi}{3}}$$

4.12 Determine whether the series converges or diverges. If it converges, find its sum.

$$10 - 6 + 3.6 - 2.16 + \cdots$$

4.12 Determine whether the series converges or diverges.

If it converges, find its sum.

$$10 - 6 + 3.6 - 2.16 + \cdots$$

The series doesn't look geometric but it could be

$$10 \cdot \boxed{r} = -6 \implies r = \frac{-6}{10} = \frac{-3}{5}$$

$$-6 \boxed{r} = 3.6 \qquad -6 \cdot \frac{-6}{10} = \frac{36}{10} = 3.6 \checkmark$$

$$3.6 \boxed{r} = -2.16 \qquad 3.6 \times \frac{-6}{10} = \frac{-21.6}{10} = -2.16 \checkmark$$

So the series is geometric with

$$r = \frac{-6}{10} \qquad |r| < 1 \implies \text{the series conv. to:}$$

$$\text{first term} = 10 \qquad \text{Sum} = \frac{10}{1 - \frac{-6}{10}} = \frac{10}{\frac{16}{10}} = \frac{100}{16}$$

$$\boxed{\text{Sum} = \frac{25}{4}}$$

4.13 Determine whether the series converges or diverges.

I. $\sum\limits_{n=1}^{\infty}\left(n-\sqrt{n^2-5n}\right)$

II. $\sum\limits_{n=1}^{\infty}(-1)^n\ln\left(1+\frac{2}{n^2}\right)$

oscillating

– converges

$$\frac{\left(n-\sqrt{n^2-5n}\right)}{1}\cdot\frac{+n+\sqrt{n^2-5n}}{+n+\sqrt{n^2-5n}}$$

$$\frac{n^2-n^2+5n}{+n-\sqrt{n^2-5n}}$$

$$\frac{5}{1+\sqrt{n^2-5n}}$$

$$\frac{5}{1+\sqrt{\frac{n^2-5n}{n^2}}}$$

$$\frac{5}{1-\sqrt{1-\frac{5n}{n^2}}^{\,0}}$$

$$\frac{5}{1+1}=\frac{5}{0}=0 \; (\text{?})$$

Diverges because

$\boxed{\neq 0}$

193

4.13 Determine whether the series converges or diverges.

I. $\sum_{n=1}^{\infty}\left(n-\sqrt{n^2-5n}\right)$ II. $\sum_{n=1}^{\infty}(-1)^n \ln\left(1+\frac{2}{n^2}\right)$

I. Test for Divergence

$$\lim_{n\to\infty} \frac{n-\sqrt{n^2-5n}}{1} \cdot \frac{n+\sqrt{n^2-5n}}{n+\sqrt{n^2-5n}} = \lim_{n\to\infty} \frac{n^2-(n^2-5n)}{n+\sqrt{n^2-5n}}$$

$$= \lim_{n\to\infty} \frac{5n}{n+\sqrt{n^2-5n}} \cdot \frac{\frac{1}{n}}{\frac{1}{n}} = \lim_{n\to\infty} \frac{5}{1+\frac{\sqrt{n^2-5n}}{n}}$$

$\sqrt{n^2}$

$\leftarrow$ rename to $\frac{1}{\sqrt{n^2}}$

$$= \lim_{n\to\infty} \frac{5}{1+\sqrt{\frac{n^2-5n}{n^2}}} = \lim_{n\to\infty} \frac{5}{1+\sqrt{1-\frac{5}{n}}} \quad \xleftarrow{\to 0 \text{ as } n\to\infty}$$

$$= \frac{5}{1+\sqrt{1}} = \boxed{\frac{5}{2}} \neq 0 .$$ The series diverges by the Test for divergence

II. Alternating Series Test

$b_n = \ln\left(1+\frac{2}{n^2}\right)$ $\xleftarrow{\to 0}$ AS $n\to\infty$ and $\ln 1 = 0$

① $\lim_{n\to\infty} b_n = \lim_{n\to\infty} \ln\left(1+\frac{2}{n^2}\right) = 0$

② $b_{n+1} < b_n$ $\ln\left(1+\frac{2}{(n+1)^2}\right) < \ln\left(1+\frac{2}{n^2}\right)$

smaller

The series converges by the Alternating Series Test

4.14 Determine whether the series converges or diverges.

I. $\displaystyle\sum_{n=1}^{\infty}\left(\ln\left(\sqrt{e}+\frac{1}{n}\right)\right)^{n+1}$

II. $\displaystyle\sum_{n=1}^{\infty}\frac{n^{\sqrt{3}}}{e^{n+4}\,5^{n}}$

$$\sqrt{\left(e+\ln\left(\tfrac{1}{n}\right)\right)^{n+1}}^{\,n}$$

4.14 Determine whether the series converges or diverges.

I. $\displaystyle\sum_{n=1}^{\infty}\left(\ln\left(\sqrt{e}+\frac{1}{n}\right)\right)^{n+1}$

II. $\displaystyle\sum_{n=1}^{\infty}\frac{n^{\sqrt{3}}}{e^{n+4}\,5^n}$

I. Root Test

$$\lim_{n\to\infty}\left(|a_n|\right)^{\frac{1}{n}}=\lim_{n\to\infty}\left[\ln\left(\sqrt{e}+\frac{1}{n}\right)^{n+1}\right]^{\frac{1}{n}}=\lim_{n\to\infty}\ln\left(\sqrt{e}+\frac{1}{n}\right)^{1+\frac{1}{n}}$$

$$\underset{\frac{1}{n}\to 0 \text{ as } n\to\infty}{}=\ln\left(\sqrt{e}+0\right)^1=\tfrac{1}{2}\ln e=\tfrac{1}{2}<1$$

The series converges by the Root Test

II. Ratio Test

$$\lim_{n\to\infty}\left|\frac{a_{n+1}}{a_n}\right|=\lim_{n\to\infty}\left|\frac{(n+1)^{\sqrt{3}}}{n^{\sqrt{3}}}\cdot\frac{e^{n+4}}{e^{n+5}}\cdot\frac{5^n}{5^{n+1}}\right|$$

$$=\lim_{n\to\infty}\left|1\cdot\frac{e^{n+4}}{e^{n+4}\cdot e}\cdot\frac{5^n}{5^n\cdot 5}\right|=\tfrac{1}{5e}<1$$

since
deg.num = deg.den.

The series converges by the Ratio Test

4.15 Determine whether the series converges or diverges.

I. $\displaystyle\sum_{n=1}^{\infty} \frac{1}{n\left(1+\left(\ln n\right)^2\right)}$ $\qquad$ II. $\displaystyle\sum_{n=1}^{\infty} \frac{1}{2\sqrt[3]{n}-\sqrt[5]{n}}$

$n + n\ln n^2$

$$\frac{\cdot 1}{(n+1)\left(1+(\ln(n+1))^2\right)} \cdot \frac{n\left(1+\ln(n)^2\right)}{1}$$

$$\frac{n}{n+1} \cdot \frac{1}{n}$$

$$\frac{1}{1+\frac{1}{n}} \quad {}^0 = 1$$

197

4.15 Determine whether the series converges or diverges.

I. $\displaystyle\sum_{n=1}^{\infty} \frac{1}{n\left(1+(\ln n)^2\right)}$
II. $\displaystyle\sum_{n=1}^{\infty} \frac{1}{2\sqrt[3]{n}-\sqrt[5]{n}}$

I. Integral Test

$f(x) = \dfrac{1}{x(1+(\ln x)^2)}$

$f(x)$ is
(A) positive
(b) Continuous
(c) decreasing

Consider

$\displaystyle\int_{1}^{\infty} f(x)\,dx$

$\displaystyle\int_{1}^{\infty} \frac{1}{x(1+(\ln x)^2)}\,dx$

$u = \ln x$
$du = \frac{1}{x}dx$

$\displaystyle\int_{0}^{\infty} \frac{1}{1+u^2}\,du = \arctan u$

$x=1 \Rightarrow u=0$
$x \to \infty \Rightarrow u \to \infty$

$= \displaystyle\lim_{b\to\infty} \arctan u \Big|_{0}^{b}$

$= \underbrace{\displaystyle\lim_{b\to\infty} \arctan b}_{\pi/2} - \underbrace{\arctan 0}_{0} = \frac{\pi}{2} \Rightarrow$ The integral converges

> The series converges by the Integral Test

II. Direct Comparison Test

$b_n = \dfrac{1}{2\sqrt[3]{n}}$ $\displaystyle\sum_{n=1}^{\infty} b_n = \frac{1}{2}\sum_{n=1}^{\infty} \frac{1}{n^{1/3}}$ Divergent p-series $p = \frac{1}{3} < 1$

$a_n = \dfrac{1}{2\sqrt[3]{n}-\sqrt[5]{n}}$ has a smaller denom (all other parts are the same)

a_n is larger

$b_n < a_n$ so $\sum a_n$ will also diverge

> The series diverges by the Direct Comparison Test

4.16 Determine whether the series converges or diverges.

I. $\displaystyle\sum_{n=1}^{\infty}\left(\frac{n+3}{2n-5}\right)^{n}$

Root

$\frac{1}{2}$ converges

$\boxed{\left|\frac{1}{2}\right| < 1}$

II. $\displaystyle\sum_{n=1}^{\infty}\sqrt[4]{\frac{2}{n^{3}}}$

$\left(\dfrac{2}{n^{3}}\right)^{\frac{1}{4}}$

$\dfrac{2^{\frac{1}{4}}}{n^{\frac{3}{4}}}$

Since p test $3/4 < 1$

$\boxed{\text{diverges}}$

4.16 Determine whether the series converges or diverges.

I. $\displaystyle\sum_{n=1}^{\infty}\left(\frac{n+3}{2n-5}\right)^{n}$

II. $\displaystyle\sum_{n=1}^{\infty}\sqrt[4]{\frac{2}{n^3}}$

I. Root Test

$$\lim_{n\to\infty}\left(|a_n|\right)^{\frac{1}{n}} = \lim_{n\to\infty}\left[\left(\frac{n+3}{2n-3}\right)^{n}\right]^{\frac{1}{n}} = \lim_{n\to\infty}\frac{n+3}{2n-3} = \frac{1}{2}<1$$

The series converges by the Root Test

II. P-series

$$\sum_{n=1}^{\infty}\sqrt[4]{\frac{2}{n^3}} = \sum_{n=1}^{\infty}\left(\frac{2}{n^3}\right)^{\frac{1}{4}} = 2^{\frac{1}{4}}\cdot\underbrace{\sum_{n=1}^{\infty}\frac{1}{n^{3/4}}}_{\text{Div. p-series}}$$

Div. p-series
$p = \frac{3}{4} < 1$

The series is a divergent p-series

4.17 Determine whether the series converges or diverges.

I. $\displaystyle\sum_{n=1}^{\infty} \frac{n^2}{\left(n^3+1\right)^3}$ $\quad \dfrac{n^2}{n^3 \cdot n} < \dfrac{n^2}{n^3 \cdot n}$

II. $\displaystyle\sum_{n=1}^{\infty} \frac{2+\sin(n)}{\sqrt{n^5}}$ $\quad$ converges

$\quad \dfrac{4}{27} \quad \dfrac{1}{2}$

$\dfrac{2}{5} \cdot \dfrac{3}{n^4}$

$\dfrac{(n+1)^2}{\left((n+1)^3+1\right)^3} \cdot \dfrac{(n^3+1)^3}{n^2} \quad \rightarrow \quad \dfrac{1}{n^2}$

p test
converges
because
$\dfrac{5}{4} > 1$

$\dfrac{2n+1}{n^2} \cdot \dfrac{1}{n^2}$

―――――――――

$n \downarrow$
$n \; n^2 \; n$
$\quad 1 \; n \; 1$

4.17 Determine whether the series converges or diverges.

I. $\displaystyle\sum_{n=1}^{\infty} \frac{n^2}{\left(n^3+1\right)^3}$

II. $\displaystyle\sum_{n=1}^{\infty} \frac{2+\sin(n)}{\sqrt{n^5}}$

I. Direct Comparison Test

$b_n = \dfrac{n^2}{n^9} = \dfrac{1}{n^7}$ $\displaystyle\sum_{n=1}^{\infty} \frac{1}{n^7}$ is a convergent p-series
$p = 7 > 1$

$a_n = \dfrac{n^2}{(n^3+1)^3}$ has a larger denom. than b_n
all other parts $=$, so we have

$a_n < b_n$ a larger denom makes a smaller fraction

The series converges by the Direct Comp. Test

II. Direct Comparison
$-1 \le \sin n \le 1$

$1 \le 2+\sin n \le 3$

$\dfrac{1}{n^{5/2}} \le \underbrace{\dfrac{2+\sin n}{n^{5/2}}}_{} \le \dfrac{3}{n^{5/2}}$

$a_n \le b_n$

$b_n = \dfrac{3}{n^{5/2}}$ $\displaystyle\sum_{n=1}^{\infty} \frac{3}{n^{5/2}}$ is a convergent p-series

The series converges by the Direct Comp. Test

202

4.18 Determine whether the series converges or diverges.

I. $\displaystyle\sum_{n=2}^{\infty} \frac{\ln n}{\sqrt{n^3}}$

II. $\displaystyle\sum_{n=1}^{\infty} \frac{\sqrt{n+4}}{n^2}$ $\dfrac{n^{\frac12}}{n^2}$ $\cdot \dfrac{\cancel{x}1}{n^{+}3}$ $\dfrac{1}{n^{\frac{3}{2}}}$

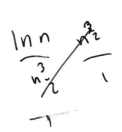

$\dfrac{\ln n}{n^{\frac{3}{2}}}$ $\dfrac{1}{n^{\frac{3}{2}}}$ ✗

$\sqrt{x}$

$\dfrac{n^{\frac12}}{n^2}$

$= n^{-\frac{3}{2}}$

converge

$\dfrac{\ln n}{n^{\frac{3}{2}}}$ $n^{\frac{3}{2}}$ $\dfrac{\ln n}{1} = \ln n \rightarrow$ $= \dfrac{1}{n^{\frac{3}{2}}}$

p test

$\dfrac{\ln n}{n^{\frac{3}{2}}}$ $\diagup$ $\dfrac{}{1}$

Direct

Limit Comparison

Ratio

Root

4.18 Determine whether the series converges or diverges.

I. $\displaystyle\sum_{n=2}^{\infty} \frac{\ln n}{\sqrt{n^3}}$ II. $\displaystyle\sum_{n=1}^{\infty} \frac{\sqrt{n+4}}{n^2}$

I. Integral Test

$f(x) = \dfrac{\ln x}{x^{3/2}}$ $f(x)$ is $\begin{cases} \text{positive} \\ \text{continuous} \\ \text{decreasing} \end{cases}$ Consider $\displaystyle\int_2 f(x)\,dx$

$n^{\frac{1}{2}} \cdot \frac{2}{1}^{\wedge}$

$\sqrt{n}$

$\displaystyle\int_2^{\infty} \frac{\ln x}{x^{3/2}}\,dx$ ← Integration by parts

$u = \ln x$ $dv = x^{-3/2}$

$du = \frac{1}{x}dx$ $v = x^{-1/2} \cdot -2$

$= \lim_{b\to\infty}\left[\frac{-\ln x - 8}{2\sqrt{x}}\right]_2^b$

$uv - \int v\,du$

$= \lim_{b\to\infty}\frac{-\ln b - 8}{2\sqrt{b}} + \frac{\ln 2 + 8}{2\sqrt{2}}$ $= \frac{-\ln x}{2\sqrt{x}} + \int 2x^{-3/2}dx$

use L'Hôpital's Rule $= \frac{-\ln x}{2\sqrt{x}} - \frac{4}{\sqrt{x}} = \frac{-\ln x - 8}{2\sqrt{x}}$

$= \lim_{b\to\infty}\frac{-\frac{1}{b}}{\frac{1}{\sqrt{b}}} = \lim_{b\to\infty}\frac{-1}{b}\cdot\sqrt{b} = \lim_{n\to\infty}\frac{-1}{\sqrt{b}} = 0$

The integral converges to $\boxed{\dfrac{\ln 2 + 8}{2\sqrt{2}}}$

n^2 $\frac{3}{2}$

$n^{\frac{1}{2}}$

$\boxed{\text{The series also converges by the Integral Test}}$

II. Limit Comparison Test

$b_n = \dfrac{\sqrt{n}}{n^2} = \dfrac{1}{n^{3/2}}$ $\sum b_n = \displaystyle\sum_{n=1}^{\infty}\frac{1}{n^{3/2}}$ convergent p-series $\left(p = \frac{3}{2} < 1\right)$

$\displaystyle\lim_{n\to\infty}\frac{a_n}{b_n} = \lim_{n\to\infty}\frac{\frac{\sqrt{n+4}}{n^2}}{\frac{1}{n^{3/2}}} = \lim_{n\to\infty}\frac{\sqrt{n+4}}{n^2}\cdot n^{3/2}$ ↑ $\sqrt{n^3}$

$= \displaystyle\lim_{n\to\infty}\frac{\sqrt{(n+4)\cdot n^3}}{n^2} = \lim_{n\to\infty}\frac{\sqrt{n^4+4n^3}}{n^2} = 1$ so they behave alike

$\boxed{\text{The series converges by the Limit Comp. Test}}$

4.19 Determine whether the series converges or diverges.

I. $\displaystyle\sum_{n=1}^{\infty} \frac{n!\,(n-1)!\cdot 5^n}{(2n+1)!}$

II. $\displaystyle\sum_{n=1}^{\infty} \left(\cos\left(\tfrac{1}{n}\right)\right)^{n^2}$

$$\frac{(n+1)(n)\,(n+1)!\,(n)!\cdot 5^{n+1}}{(2n+3)!} \cdot \frac{(2n+1)!}{n!\,(n)!\cdot 5^n}$$

$\dfrac{5}{(2n+3)(2n+2)}$

$\dfrac{(2n+1)!}{(2n+3)(2n+2)}$

$$\frac{5(n+1)(n)}{(2n+1)(2n+2)}$$

(II)

$\displaystyle\lim_{b\to\infty}\sum_{n=1}^{b} \cos$

$\dfrac{1}{n}$

$\dfrac{n^2}{5^n}$ (5)

$\displaystyle\lim_{b\to\infty}\sum_{n=1}^{b}\sqrt[n]{(\cos\tfrac{1}{n})^{n^2}}$

$(\cos\tfrac{1}{n})^n$

205

4.19 Determine whether the series converges or diverges.

I. $\displaystyle\sum_{n=1}^{\infty} \frac{n!\cdot(n-1)!\cdot 5^n}{(2n+1)!}$ II. $\displaystyle\sum_{n=1}^{\infty}\left(\cos\left(\frac{1}{n}\right)\right)^{n^2}$

I. Ratio Test

$$\lim_{n\to\infty}\left|\frac{a_{n+1}}{a_n}\right| = \lim_{n\to\infty}\left|\frac{(n+1)!}{n!}\cdot\frac{n!}{(n-1)!}\cdot\frac{5^{n+1}}{5^n}\cdot\frac{(2n+1)!}{(2n+3)!}\right|$$

$$= \lim_{n\to\infty}\left|\frac{(n+1)n!}{n!}\cdot\frac{n\cdot(n-1)!}{(n-1)!}\cdot\frac{5^n\cdot5}{5^n}\cdot\frac{(2n+1)!}{(2n+3)(2n+2)(2n+1)!}\right|$$

$$= \lim_{n\to\infty}\frac{(n+1)\cdot n\cdot 5}{(2n+3)(2n+2)} = \lim_{n\to\infty}\frac{5n^2+5n}{4n^2+10n+6} = \frac{5}{4}>1$$

$\boxed{\text{The series diverges by the Ratio Test}}$

II. Test for Divergence

$$\lim_{n\to\infty}\left(\cos\left(\frac{1}{n}\right)\right)^{n^2} = "1^\infty" \leftarrow\text{indeterminant power}$$

$$y = \lim_{n\to\infty}\left(\cos\left(\frac{1}{n}\right)\right)^{n^2}\quad\text{take } \ln$$

$$\ln y = \lim_{n\to\infty}\ln\left(\cos\left(\frac{1}{n}\right)\right)^{n^2} = \lim_{n\to\infty}n^2\ln\left(\cos\left(\frac{1}{n}\right)\right) = "\infty\cdot0"$$

$$\ln y = \lim_{n\to\infty}\frac{\ln\left(\cos\left(\frac{1}{n}\right)\right)}{\frac{1}{n^2}} = "\frac{0}{0}"\quad\text{use L'Hôpital's Rule}$$

$$\ln y \overset{L'H}{=} \lim_{n\to\infty}\frac{\frac{1}{\cos(1/n)}\cdot-\sin\left(\frac{1}{n}\right)\cdot\frac{-1}{n^2}}{\frac{-2}{n^3}} = \lim_{n\to\infty}\frac{-\tan\left(\frac{1}{n}\right)\cdot\frac{-1}{n^2}\cdot\frac{n^3}{-2}}{\frac{n}{2}}$$

$$\ln y = \lim_{n\to\infty}\frac{-\tan\left(\frac{1}{n}\right)}{\frac{2}{n}} = "\frac{0}{0}" \overset{L'H}{=} \lim_{n\to\infty}\frac{-\sec^2\left(\frac{1}{n}\right)\cdot\frac{-1}{n^2}}{\frac{-2}{n^2}}$$

$$\ln y = \lim_{n\to\infty}\frac{-\sec^2\left(\frac{1}{n}\right)}{2} = -\frac{1}{2}\qquad e^{\ln y}=e^{\frac{-1}{2}}\quad y=e^{-1/2}=\frac{1}{\sqrt{e}}\neq 0$$

$\boxed{\text{The series diverges by the Test for Divergence}}$

4.20 Determine whether the following series is absolutely convergent, conditionally convergent or divergent.

I. $\displaystyle\sum_{n=2}^{\infty} \frac{(-1)^n \ln n}{n^{3/4}}$

II. $\displaystyle\sum_{n=1}^{\infty} \frac{(n+2)!}{e^{n^2}}$

4.20 Determine whether the following series is absolutely convergent, conditionally convergent or divergent.

I. $\sum_{n=2}^{\infty} \dfrac{(-1)^n \ln n}{n^{3/4}}$

II. $\sum_{n=1}^{\infty} \dfrac{(n+2)!}{e^{n^2}}$

I. $\sum_{n=2}^{\infty} |a_n| = \sum_{n=2}^{\infty} \dfrac{\ln n}{n^{3/4}}$. $\underline{\text{Diverges}}$ by the Direct Comp. Test

$\quad$ with $b_n = \dfrac{1}{n^{3/4}}$ $\sum b_n$ is a divergent p-series

$\qquad b_n < a_n$ since $1 < \ln n$ for $n \geq 3$

$\sum_{n=2}^{\infty} \dfrac{(-1)^n \ln n}{n^{3/4}}$ converges by the $\underline{\text{Alternating Series Test}}$

$\quad b_n = \dfrac{\ln n}{n^{3/4}}$ ① $\lim_{n\to\infty} \dfrac{\ln n}{n^{3/4}} = 0$ $\quad n^{3/4}$ grows faster

$\qquad$ ② $\dfrac{\ln(n+1)}{(n+1)^{3/4}} < \dfrac{\ln n}{n^{3/4}}$ den. grows faster than num.

$\boxed{\text{The series is conditionally convergent}}$

II. The series isn't alternating so it is either absolutely convergent or divergent.

$\quad$ Use Ratio Test $\qquad\qquad\qquad\qquad (n+1)^2 = n^2 + 2n + 1$

$\lim_{n\to\infty} \left| \dfrac{a_{n+1}}{a_n} \right| = \lim_{n\to\infty} \left| \dfrac{(n+3)!}{(n+2)!} \cdot \dfrac{e^{n^2}}{e^{(n+1)^2}} \right|$

$\qquad = \lim_{n\to\infty} \left| \dfrac{(n+3)(n+2)!}{(n+2)!} \cdot \dfrac{e^{n^2}}{e^{n^2} \cdot e^{2n+1}} \right|$

$\qquad = \lim_{n\to\infty} \dfrac{n+3}{e^{2n+1}} = 0 < 1$ $\quad e^{2n+1}$ grows faster than $n+3$

$\boxed{\text{The series is absolutely convergent}}$

4.21 Determine whether the following series is absolutely convergent, conditionally convergent or divergent.

I. $\displaystyle\sum_{n=1}^{\infty}\frac{(-1)^{n}}{n+\ln n}$

II. $\displaystyle\sum_{n=1}^{\infty}\frac{(-1)^{n-1}(n!)^{3}}{(3n)!}$

$< \dfrac{1}{n+\ln n}$

$\displaystyle\sum_{n=1}^{\infty}$

$\dfrac{(n!)^{3}}{(3n)!}$

$\dfrac{(n+1)^{3}!}{(3(n+1))!} \cdot \dfrac{(3n)!}{(n!)^{3}}$

Direct Comp
Root test
Ratio test
Limit comparison

$(-1)^{n-1}$

$\dfrac{(n+1)6! \quad (n+1)! \quad (n+1)!}{n \qquad n \qquad n} \cdot \dfrac{(3n!)}{(3n+3)!}$

$(3n+2)(3n+1)$

4.21 Determine whether the following series is absolutely convergent, conditionally convergent or divergent.

I. $\displaystyle\sum_{n=1}^{\infty} \frac{(-1)^n}{n+\ln n}$

II. $\displaystyle\sum_{n=1}^{\infty} \frac{(-1)^{n-1}(n!)^3}{(3n)!}$

I. $\displaystyle\sum_{n=1}^{\infty}|a_n| = \sum_{n=1}^{\infty}\frac{1}{n+\ln n}$ Use Limit comp. test

$b_n = \frac{1}{n}$ $\sum\frac{1}{n}$ div. Harmonic

$\displaystyle\lim_{n\to\infty}\frac{a_n}{b_n} = \lim_{n\to\infty}\frac{\frac{1}{n+\ln n}}{\frac{1}{n}} = \lim_{n\to\infty}\frac{n}{n+\ln n} = 1$

$\ln n$ negligible when $n\to\infty$

So they behave alike and $\sum|a_n|$ diverges.

$\displaystyle\sum_{n=1}^{\infty}\frac{(-1)^n}{n+\ln n}$ use Alternating Series Test

$b_n = \frac{1}{n+\ln n}$

①$\displaystyle\lim_{n\to\infty}\frac{1}{n+\ln n} = 0$

②$b_{n+1} < b_n$

$\Bigg\}$ the series converges by the AST

$\boxed{\text{The series is conditionally convergent}}$

II. $\displaystyle\sum|a_n| = \sum_{n=1}^{\infty}\frac{n!\cdot n!\cdot n!}{(3n)!}$ use Ratio Test

$\displaystyle\lim_{n\to\infty}\left|\frac{a_{n+1}}{a_n}\right| = \lim_{n\to\infty}\left|\frac{(n+1)!}{n!}\cdot\frac{(n+1)!}{n!}\cdot\frac{(n+1)!}{n!}\cdot\frac{(3n)!}{(3n+3)!}\right|$

$= \displaystyle\lim_{n\to\infty}\left|\frac{(n+1)\cdot n!}{n!}\cdot\frac{(n+1)\cdot n!}{n!}\cdot\frac{(n+1)\cdot n!}{n!}\cdot\frac{(3n)!}{(3n+3)(3n+2)(3n+1)(3n)!}\right|$

$= \displaystyle\lim_{n\to\infty}\frac{(n+1)^3}{(3n+3)(3n+2)(3n+1)} = \frac{1}{27} < 1$

deg num = deg den.

$\boxed{\text{The series converges absolutely}}$

4.22 Determine whether the following series is absolutely convergent, conditionally convergent or divergent.

I. $\displaystyle\sum_{n=1}^{\infty} \frac{(-e)^n}{(2.71)^n}$ $\dfrac{(-1)^n (e^n)}{\overline{2.71^n}}$

II. $\displaystyle\sum_{n=1}^{\infty} \frac{1}{\sqrt{n}\left(\sqrt{n}+1\right)^3}$

$\dfrac{e^n}{(2.71)^n} \overset{no}{<} \dfrac{1}{(2.71)^n} \to$ converges

a $f(n+1) < f(n)$

$\dfrac{(-e)^{n+1}}{(2.71)^{n+1}}$

Denom
bigger

4.22 Determine whether the following series is absolutely convergent, conditionally convergent or divergent.

I. $\displaystyle\sum_{n=1}^{\infty} \frac{(-e)^n}{(2.71)^n}$ $-e$

$\overline{2.71}$

II. $\displaystyle\sum_{n=1}^{\infty} \frac{1}{\sqrt{n}\left(\sqrt{n}+1\right)^3}$

I. $\displaystyle\sum |a_n| = \sum_{n=1}^{\infty}\left(\frac{e}{2.71}\right)^n$ Geometric Series

with $r = \dfrac{e}{2.71} > 1$

Diverges

$e \approx 2.7182$

$e > 2.71$

$\sum a_n$ will also diverge   now $r = \dfrac{-e}{2.71}$

$|r| \geq 1$

$\boxed{\text{The series diverges}}$

II. The series is not alternating so it either
Converges absolutely or diverges.

Direct compare to $b_n = \dfrac{1}{\sqrt{n}\cdot\sqrt{n}^3} = \dfrac{1}{n^2}$

$\sum \dfrac{1}{n^2}$ converges p-series $p > 1$

$a_n < b_n$

$\dfrac{1}{\sqrt{n}(\sqrt{n}+1)^3} < \dfrac{1}{\sqrt{n}(\sqrt{n})^3}$

larger den.
so smaller

$\boxed{\text{The series is absolutely convergent}}$

4.23 Determine whether the following series is absolutely convergent, conditionally convergent or divergent.

I. $\displaystyle\sum_{n=1}^{\infty}\frac{(-1)^{n+1}}{n\sqrt{n+2}}$

II. $\displaystyle\sum_{n=2}^{\infty}\frac{(-1)^{n}\,n^{2/3}}{\ln n}$

4.23 Determine whether the following series is absolutely convergent, conditionally convergent or divergent.

I. $\sum_{n=1}^{\infty} \dfrac{(-1)^{n+1}}{n\sqrt{n+2}}$

II. $\sum_{n=2}^{\infty} \dfrac{(-1)^{n} n^{2/3}}{\ln n}$

I. $\sum_{n=1}^{\infty} |a_n| = \sum_{n=1}^{\infty} \dfrac{1}{n\sqrt{n+2}}$ Direct compare to $b_n = \dfrac{1}{n\sqrt{n}}$

$\sum \dfrac{1}{n\sqrt{n}}$ conv. p-series $p = \dfrac{3}{2} > 1$

$a_n < b_n$

$\dfrac{1}{n\sqrt{n+2}} < \dfrac{1}{n\sqrt{n}}$ larger denom $\rightarrow$ smaller fraction

all other parts same

$\boxed{\text{The series is absolutely convergent}}$

II. $\sum |a_n| = \sum_{n=2}^{\infty} \dfrac{n^{2/3}}{\ln n}$ $\lim\limits_{n\to\infty} \dfrac{n^{2/3}}{\ln n} = \infty$

Diverges by the Test for Div. $n^{2/3}$ grows faster than $\ln n$

$\sum a_n$ will still diverge $\lim\limits_{n\to\infty} \dfrac{(-1)^n \cdot n^{2/3}}{\ln n}$ DNE

$\begin{cases} \infty & n \text{ is even} \\ -\infty & n \text{ is odd} \end{cases}$

$\boxed{\text{The series diverges}}$

a) $\left(\frac{-3}{2}, \frac{1}{2}\right]$ e) $(1,2]$

b) $\left[\frac{-3}{2}, \frac{1}{2}\right]$ f) $[1,2]$

c) $\left(\frac{-3}{2}, \frac{1}{2}\right)$ g) $\left\{\frac{-1}{2}\right\}$

4.24 Find the interval of convergence of the power series.

d) $\left[\frac{-3}{2}, \frac{1}{2}\right)$ h) $(-\infty, \infty)$

$$\sum_{n=1}^{\infty} \frac{(n+1)(2x+1)^n}{2^n n^2}$$

a) $\left(\frac{-3}{2},\frac{1}{2}\right]$ e) $(1,2]$

b) $\left[\frac{-3}{2},\frac{1}{2}\right]$ f) $[1,2]$

c) $\left(\frac{-3}{2},\frac{1}{2}\right)$ g) $\left\{\frac{-1}{2}\right\}$

4.24 Find the interval of convergence of the power series.

d) $\left[\frac{-3}{2},\frac{1}{2}\right)$ h) $(-\infty,\infty)$

$$\sum_{n=1}^{\infty}\frac{(n+1)(2x+1)^n}{2^n n^2}$$

$$\lim_{h\to\infty}\left|\frac{a_{n+1}}{a_n}\right| = \lim_{h\to\infty}\left|\frac{\overbrace{n+2}^{(n+1)\cdot 1}}{n+1}\cdot\frac{(2x+1)^{n+1}}{(2x+1)^n}\cdot\frac{2^n}{2^{n+1}}\cdot\frac{n^2}{(n+1)^2}\right|$$

$$= \lim_{h\to\infty}\left|1\cdot\frac{(2x+1)\cdot(2x+1)}{(2x+1)}\cdot\frac{2^n}{2^n\cdot 2}\cdot 1\right|$$

$$\left|\frac{2x+1}{2}\right| \overset{force}{<} 1$$

$$\Rightarrow |2x+1| < 2 \qquad -2 < 2x+1 < 2$$
$$\qquad\qquad\qquad\qquad -1 \qquad -1 \quad -1$$
$$\qquad\qquad\qquad\qquad \overline{-3 < 2x < 1}$$
$$\qquad\qquad\qquad\qquad \frac{-3}{2} \quad \frac{2x}{2} \quad \frac{1}{2}$$
$$\qquad\qquad\qquad\qquad \boxed{\tfrac{-3}{2}} < x < \boxed{\tfrac{1}{2}}$$

$\underline{x=\frac{-3}{2}}$

$$\sum_{n=1}^{\infty}\frac{(n+1)(-2)^n}{2^n\cdot n^2} = \sum_{n=1}^{\infty}\frac{(-1)^n\cdot(n+1)}{n^2}$$

$\boxed{Conv.}$ by AST

$b_n = \frac{n+1}{n^2} = \frac{1}{n}+\frac{1}{n^2}$

$\lim_{h\to\infty}b_n = 0$ / b_n is decr.

$\underline{x=\frac{1}{2}}$

$$\sum_{n=1}^{\infty}\frac{(n+1)2^n}{n^2\cdot 2^n}$$

$$\sum_{n=1}^{\infty}\frac{n+1}{n^2} = \sum_{n=1}^{\infty}\frac{1}{n}+\frac{1}{n^2}$$

$$\underset{\underset{DIV}{n=1}}{\sum}\frac{1}{n} + \underset{\underset{CONV.}{n=1}}{\sum}\frac{1}{n^2} \quad\Big\}\boxed{DIV}$$

$$\boxed{\left[\frac{-3}{2},\frac{1}{2}\right)}$$

(A) $(-1, 2]$ (E) $(1, 2]$

(B) $(-1, 2)$ (F) $[1, 2]$

(C) $(1, 2)$ (G) $[-1, 2]$

(D) $[1, 2)$ (H) $(-\infty, \infty)$

4.25 Find the interval of convergence of the power series.

$$\sum_{n=1}^{\infty} \frac{(-1)^n (2x-3)^n}{2n+1}$$

4.25 Find the interval of convergence of the power series.

$$\sum_{n=1}^{\infty} \frac{(-1)^n (2x-3)^n}{2n+1}$$

$$\lim_{n \to \infty} \left| \frac{a_{n+1}}{a_n} \right| = \lim_{n \to \infty} \left| \frac{(-1)^{n+1}}{(-1)^n} \cdot \frac{(2x-3)^{n+1}}{(2x-3)^n} \cdot \frac{2n+1}{2n+3} \right|$$

$$= \lim_{n \to \infty} \left| \frac{(-1)^n (-1)}{(-1)^n} \cdot \frac{(2x-3)^n (2x-3)}{(2x-3)^n} \cdot \frac{2n+1}{2n+3} \right|$$

force → $\boxed{1}$

$$|-1 \cdot (2x-3)| < 1$$

$$|-1| \cdot |2x-3| < 1$$

$$\underbrace{|-1|}_{1} \cdot |2x-3| < 1$$

$-1 < 2x-3 < 1$

$+3 \quad +3 \quad +3$

$$\frac{2}{2} < \frac{2x}{2} < \frac{4}{2}$$

$1 < x < 2$

$x = 1$

$$\sum_{n=1}^{\infty} \frac{(-1)^n \cdot (-1)^n}{2n+1} = \sum_{n=1}^{\infty} \frac{1}{2n+1}$$

LCT with $b_n = \frac{1}{2n}$

$\sum \frac{1}{2n}$ div. (Harmonic)

$$\lim_{n \to \infty} \frac{a_n}{b_n} = \frac{\frac{1}{2n+1}}{\frac{1}{2n}} = 1$$

So they behave alike

$\boxed{\text{Div.}}$

$x = 2$

$$\sum_{n=1}^{\infty} \frac{(-1)^n}{2n+1} \quad \boxed{\text{Conv}} \text{ by AST}$$

with $b_n = \frac{1}{2n+1}$

1) $\lim_{n \to \infty} b_n = 0$

2) b_n is decreasing

$$\boxed{(1, 2]}$$

4.26 Find the interval of convergence of the power series.

$$\sum_{n=1}^{\infty} \frac{(2x-3)^n}{5^n \cdot \sqrt[4]{n^3}}$$

A) $(-4, 4]$ C) $[-1, 4)$ E) $(-1, 4)$

B) $[-1, 4]$ D) $(1, 4]$ F) $[-4, 1)$

4.26 Find the interval of convergence of the power series.

$$\sum_{n=1}^{\infty} \frac{(2x-3)^n}{5^n \cdot \sqrt[4]{n^3}}$$

A) $(-4, 4]$ C) $[-1, 4)$ E) $(-1, 4)$

B) $[-1, 4]$ D) $(1, 4]$ F) $[-4, 1)$

$$\lim_{n \to \infty} \left| \frac{a_{n+1}}{a_n} \right| = \lim_{n \to \infty} \left| \frac{(2x-3)^{n+1}}{(2x-3)^n} \cdot \frac{5^n}{5^{n+1}} \cdot \frac{\sqrt[4]{n^3}}{\sqrt[4]{(n+1)^3}} \right|$$

$$= \lim_{n \to \infty} \left| \frac{(2x-3)^n \cdot (2x-3)}{(2x-3)^n} \cdot \frac{5^n}{5^n \cdot 5} \cdot 1 \right|$$

$$\left| \frac{2x-3}{5} \right| < 1 \quad \text{force}$$

$$|2x-3| < 5$$

$$-5 < 2x-3 < 5$$
$$+3 \qquad +3 \quad +3$$

$$\overline{\frac{-2}{2} < \frac{2x}{2} < \frac{8}{2}}$$

$$-1 < x < 4$$

$$\underline{x = -1}$$

$$\sum_{n=1}^{\infty} \frac{(-5)^n}{5^n \cdot n^{3/4}} = \sum_{n=1}^{\infty} \frac{(-1)^n}{n^{3/4}} \quad \boxed{\text{Conv}} \quad \text{AST } b_n = \frac{1}{n^{3/2}}$$

① b_n decr.

② $\lim_{n \to \infty} b_n = 0$

$$\underline{x = 4}$$

$$\sum_{n=1}^{\infty} \frac{5^n}{5^n \cdot n^{3/4}} = \sum_{n=1}^{\infty} \frac{1}{n^{3/4}} \quad \boxed{\text{DIV}} \quad p\text{-series } p = \frac{3}{4}$$

$$\boxed{[-1, 4)}$$

4.27 Find the coefficient on x^4 in the Maclaurin series for
$$f(x) = xe^{-2x}$$

A) $\dfrac{2}{3}$ B) $\dfrac{4}{3}$ C) $\dfrac{-8}{3}$ D) $-\dfrac{2}{3}$

E) $-\dfrac{4}{3}$ F) -2 G) 2 H) -1

4.27 Find the coefficient on x^4 in the Maclaurin series for

$$f(x) = xe^{-2x}$$

A) $\dfrac{2}{3}$ B) $\dfrac{4}{3}$ C) $\dfrac{-8}{3}$ D) $-\dfrac{2}{3}$

E) $-\dfrac{4}{3}$ F) -2 G) 2 H) -1

$$e^x = 1 + x + \frac{x^2}{2!} + \frac{x^3}{3!} + \frac{x^4}{4!}$$

$$e^{-2x} = 1 - 2x + \frac{(-2x)^2}{2!} + \frac{(-2x)^3}{3!} + \frac{(-2x)^4}{4!} + \cdots$$

$$xe^{-2x} = x - 2x^2 + \frac{4x^3}{2} - \frac{8x^4}{6} + \cdots$$

$$\text{Coeff. on } x^4 = \boxed{\dfrac{-4}{3}}$$

a) 1
e) $\dfrac{1}{24}$

4.28 Find the coefficient on x^6 in the Maclaurin series for $f(x) = \cosh(2x)$ using the identity $\cosh(x) = \dfrac{1}{2}\left(e^x + e^{-x}\right)$

b) $\dfrac{4}{45}$
f) $\dfrac{1}{720}$
c) $\dfrac{8}{35}$
d) 2
h) $\dfrac{2}{315}$

a) 1   e) $\dfrac{1}{24}$

4.28 Find the coefficient on x^6 in the Maclaurin series for

b) $\dfrac{1}{45}$ f) $\dfrac{1}{720}$

$f(x) = \cosh(2x)$ using the identity $\cosh(x) = \dfrac{1}{2}\left(e^x + e^{-x}\right)$

c) $\dfrac{1}{3}$

d) 2   h) $\dfrac{2}{315}$

$$e^x = 1 + x + \frac{x^2}{2!} + \frac{x^3}{3!} + \frac{x^4}{4!} + \frac{x^5}{5!} + \frac{x^6}{6!} + \cdots$$

$$+ \; e^{-x} = 1 - x + \frac{x^2}{2!} - \frac{x^3}{3!} + \frac{x^4}{4!} - \frac{x^5}{5!} + \frac{x^6}{6!}$$

$$e^x + e^{-x} = 2 + 2\left(\frac{x^2}{2!}\right) + 2\left(\frac{x^4}{4!}\right) + 2\left(\frac{x^6}{6!}\right) + \cdots$$

$$\frac{e^x + e^{-x}}{2} = 1 + \frac{x^2}{2!} + \frac{x^4}{4!} + \frac{x^6}{6!} + \cdots = \cosh(x)$$

$$\cosh(2x) = 1 + \frac{(2x)^2}{2!} + \frac{(2x)^4}{4!} + \frac{(2x)^6}{6!} + \cdots$$

$$\text{Coeff. on } x^6 = \frac{2^6}{6!} = \frac{64}{720} = \frac{8}{90} = \boxed{\frac{4}{45}}$$

224

b) $\dfrac{1}{24}$ f) $\dfrac{11!}{4!}$

c) $\dfrac{121}{16}$ g) $\dfrac{11}{5}$

d) $\dfrac{13}{6}$ h) $\dfrac{11}{4}$

4.29 If $f(x) = x^3 \cos(x^2)$, find $f^{(11)}(0)$, the value of the eleventh derivative evaluated at 0.

5!

b) $\dfrac{1}{24}$ f) $\dfrac{11!}{4!}$

c) $\dfrac{121}{16}$ g) $\dfrac{11}{5}$

d) $\dfrac{13}{6}$ h) $\dfrac{11}{4}$

4.29 If $f(x) = x^3 \cos(x^2)$, find $f^{(11)}(0)$, the value of the eleventh derivative evaluated at 0.

$$\cos x = 1 - \frac{x^2}{2!} + \frac{x^4}{4!} - \frac{x^6}{6!} + \cdots$$

$$\cos(x^2) = 1 - \frac{(x^2)^2}{2!} + \frac{(x^2)^4}{4!} - \frac{(x^2)^6}{6!} + \cdots$$

$$\cos(x^2) = 1 - \frac{x^4}{2!} + \frac{x^8}{4!} - \frac{x^{12}}{6!} + \cdots$$

$$x^3 \cdot \cos(x^2) = x^3 - \frac{x^7}{2!} + \frac{x^{11}}{4!} - \frac{x^{15}}{6!} + \cdots$$

$$\boxed{\text{Coeff. on } x^{11} = \frac{1}{4!}}$$

Maclaurin series in general:

$$f(0) + f'(0)x + \frac{f''(0)}{2!}x^2 + \frac{f'''(0)}{3!}x^3 + \cdots + \frac{f^{(11)}(0)}{11!}x^{11} + \cdots$$

$$\boxed{\text{Coeff. on } x^{11} = \frac{f^{(11)}(0)}{11!}}$$

They must equal: $\dfrac{1}{4!} = \dfrac{f^{(11)}(0)}{11!}$

$$\Rightarrow \boxed{f^{(11)}(0) = \frac{11!}{4!}}$$

226

4.30 Find the third degree Taylor polynomial for $f(x) = x^{3/2}$ centered at $x = 4$.

4.30 Find the third degree Taylor polynomial for $f(x) = x^{3/2}$ centered at $x = 4$.

$f(x) = x^{3/2}$

$f'(x) = \frac{3}{2} x^{1/2}$

$f''(x) = \frac{3}{4} \cdot x^{-1/2}$

$f'''(x) = -\frac{3}{8} x^{-3/2}$

$f(4) = 4^{3/2} = 8 \div 0! = 8$

$f'(4) = \frac{3}{2}\sqrt{4} = 3 \div 1! = 3(x-4)$

$f''(4) = \frac{3}{4} \cdot \frac{1}{\sqrt{4}} = \frac{3}{8} \div 2! = \frac{3}{16}(x-4)^2$

$f'''(4) = -\frac{3}{8} \cdot \frac{1}{\frac{4^{3/2}}{8}} = \frac{-3}{64} \div 3! = \frac{-3}{64 \cdot 6} = \frac{-1}{128}(x-4)^3$

$$T_3(x) = \frac{-1}{128}(x-4)^3 + \frac{3}{16}(x-4)^2 + 3(x-4) + 8$$

4.31 Find the Taylor polynomial of degree 2 for $f(x) = \arcsin x$ centered at $x = \dfrac{1}{2}$.

What is the quadratic coefficient? $\left(\text{Find the coefficient on } \left(x - \tfrac{1}{2}\right)^2\right)$

(A) $\dfrac{1}{3}$

(B) $\dfrac{2}{3\sqrt{3}}$

(C) $\dfrac{2}{\sqrt{3}}$

(D) $\dfrac{3}{4\sqrt{3}}$

(E) $\dfrac{1}{\sqrt{3}}$

(F) $\dfrac{\pi}{6}$

4.31 Find the Taylor polynomial of degree 2 for $f(x) = \arcsin x$ centered at $x = \dfrac{1}{2}$.

What is the quadratic coefficient? $\left(\text{Find the coefficient on } \left(x - \tfrac{1}{2}\right)^2\right)$

(A) $\dfrac{1}{3}$ (B) $\dfrac{2}{3\sqrt{3}}$ (C) $\dfrac{2}{\sqrt{3}}$

(D) $\dfrac{3}{4\sqrt{3}}$ (E) $\dfrac{1}{\sqrt{3}}$ (F) $\dfrac{\pi}{6}$

$f(x) = \arcsin x$

$f'(x) = \dfrac{1}{\sqrt{1-x^2}} = (1-x^2)^{-1/2}$

$f''(x) = -\tfrac{1}{2}(1-x^2)^{-3/2} \cdot (-2x) = \dfrac{x}{(1-x^2)^{3/2}}$

$f(\tfrac{1}{2}) = \arcsin\tfrac{1}{2}$

$f'(\tfrac{1}{2}) = \dfrac{1}{\sqrt{1-\frac{1}{4}}}$

$f''(\tfrac{1}{2}) = \dfrac{\tfrac{1}{2}}{(1-\frac{1}{4})^{3/2}}$

$\arcsin\tfrac{1}{2} = \dfrac{\pi}{6}$

$\dfrac{1}{\sqrt{\frac{3}{4}}} = \dfrac{1}{\frac{\sqrt{3}}{2}} = \dfrac{2}{\sqrt{3}}$

$\div 0! = \pi/6$

$\div 1! = \dfrac{2}{\sqrt{3}}$

$\dfrac{\frac{1}{2}}{(\frac{3}{4})^{3/2}} = \dfrac{\frac{1}{2}}{(\frac{\sqrt{3}}{2})^3} = \tfrac{1}{2} \cdot \dfrac{8}{3\sqrt{3}} = \dfrac{4}{3\sqrt{3}} \div 2! = \dfrac{2}{3\sqrt{3}}$

$\dfrac{\pi}{6} + \dfrac{2}{\sqrt{3}}(x-\tfrac{1}{2}) + \dfrac{2}{3\sqrt{3}}(x-\tfrac{1}{2})^2$

Ans $\boxed{\dfrac{2}{3\sqrt{3}}}$

4.32 Evaluate the limit
$$\lim_{x \to 0} \frac{x^2 + x \ln(1-x)}{xe^{-5x} - x + 5x^2}.$$

b) $\dfrac{1}{4}$

c) $\dfrac{-1}{100}$

d) $\dfrac{-1}{15}$

f) $\dfrac{-1}{5}$

g) $\dfrac{-1}{50}$

h) $\dfrac{-1}{25}$

b) $\dfrac{1}{4}$ f) $\dfrac{-1}{5}$

c) $\dfrac{-1}{100}$ g) $\dfrac{-1}{50}$

4.32 Evaluate the limit d) $\dfrac{-1}{15}$ h) $\dfrac{-1}{25}$

$$\lim_{x \to 0} \frac{x^2 + x\ln(1-x)}{xe^{-5x} - x + 5x^2}.$$

$$\ln(1-x) = -x - \frac{x^2}{2} - \frac{x^3}{3} - \cdots$$

$$e^x = 1 + x + \frac{x^2}{2!} + \frac{x^3}{3!} + \cdots$$

$$e^{-5x} = 1 - 5x + \frac{25x^2}{2} - \frac{125x^3}{6} + \cdots$$

$$\lim_{x \to 0} \frac{x^2 + x\left(-x - \frac{x^2}{2} - \frac{x^3}{3}\right)}{x\left(1 - 5x + \frac{25x^2}{2} - \frac{125x^3}{6} + \cdots\right) - x + 5x^2}$$

$$= \lim_{x \to 0} \frac{\cancel{x^2} - \cancel{x^2} - \frac{x^3}{2} - \frac{x^4}{3} \cdots}{\left(\cancel{x} - 5x^2 + \frac{25}{2}x^3 + \cdots\right) - \cancel{x} + 5x^2}$$

$$= \lim_{x \to 0} \frac{-\frac{x^3}{2} - \frac{x^4}{3} - \cdots}{\frac{25}{2}x^3 - \frac{125}{6}x^4 + \cdots} \cdot \frac{\frac{1}{x^3}}{\frac{1}{x^3}}$$

$$= \lim_{x \to 0} \frac{-\frac{1}{2} - \frac{1}{3}x + \text{higher order terms}^{\to 0 \text{ as } x \to 0}}{\frac{25}{2} - \frac{125}{6}x + \text{higher order terms}}$$

$$= \frac{-\frac{1}{2}}{\frac{25}{2}} = -\frac{1}{2} \cdot \frac{2}{25} = \boxed{\frac{-1}{25}}$$

232

4.33 Evaluate the limit

$$\lim_{x \to 0} \frac{\dfrac{x^2}{(1-x)^2}}{4x - 1 + e^{-4x}}$$

A) $\dfrac{1}{8}$ B) $\dfrac{1}{16}$ C) $\dfrac{1}{4}$ D) 8

E) $\dfrac{1}{2}$ F) 1   G) 2   H) 16

4.33 Evaluate the limit

$$\lim_{x\to 0} \frac{\dfrac{x^2}{(1-x)^2}}{4x-1+e^{-4x}}$$

A) $\dfrac{1}{8}$ B) $\dfrac{1}{16}$ C) $\dfrac{1}{4}$ D) 8

E) $\dfrac{1}{2}$ F) 1   G) 2   H) 16

$$\frac{1}{(1-x)^2} = 1 + 2x + 3x^2 + 4x^3 + \cdots$$

$$\frac{x^2}{(1-x)^2} = x^2 + 2x^3 + 3x^4 + 4x^5$$

$$e^x = 1 + x + \frac{x^2}{2!} + \frac{x^3}{3!} + \cdots$$

$$e^{-4x} = 1 - 4x + \frac{(-4x)^2}{2} + \frac{(-4x)^3}{3!} + \cdots$$

$$\lim_{x\to 0} \frac{x^2 + 2x^3 + 3x^4 + 4x^5}{4x - 1 + \left(1 - 4x + 8x^2 - \frac{64x^3}{6} + \cdots\right)}$$

$$= \lim_{x\to 0} \frac{\left(x^2 + 2x^3 + 3x^4 + 4x^5 + \cdots\right)\frac{1}{x^2}}{\left(8x^2 - \frac{32}{3}x^3 + \cdots\right)\frac{1}{x^2}}$$

$$= \lim_{x\to 0} \frac{1 + 2x + 3x^2 + 4x^3 + \cdots \to 0}{8 - \frac{32}{3}x + \cdots \to 0} = \boxed{\frac{1}{8}}$$

234

4.34 Find the sum of the series

$$\ln\left(1+2+\frac{2^2}{2!}+\frac{2^3}{3!}+\cdots\right)+\left(1+\frac{2}{3}+\frac{4}{9}+\frac{8}{27}+\cdots\right)$$

A) 5   B) 4   C) 3   D) 2   E) 1   F) diverges

4.34 Find the sum of the series

$$\ln\left(1+2+\frac{2^2}{2!}+\frac{2^3}{3!}+\cdots\right)+\left(1+\frac{2}{3}+\frac{4}{9}+\frac{8}{27}+\cdots\right)$$

A) 5   B) 4   C) 3   D) 2   E) 1   F) diverges

$$\ln\left(1+2+\frac{2^2}{2!}+\frac{2^3}{3!}+\cdots\right)+\left(1+\frac{2}{3}+\frac{4}{9}+\frac{8}{27}+\cdots\right)$$

e^x power series w/
x replaced by 2

Geometric w/ $r = \frac{2}{3}$
first term 1

$$\ln(e^2) + \frac{1}{1-\frac{2}{3}}$$

$$2 + \frac{1}{\frac{1}{3}} = 2+3 = \boxed{5}$$

236

4.35 In order to find the sum of the series $\displaystyle\sum_{n=1}^{\infty}\frac{1}{n^5}$ correct to 3 decimal places, how many terms should be used?

4.35 In order to find the sum of the series $\sum\limits_{n=1}^{\infty} \dfrac{1}{n^5}$

correct to 3 decimal places, how many terms should be used?

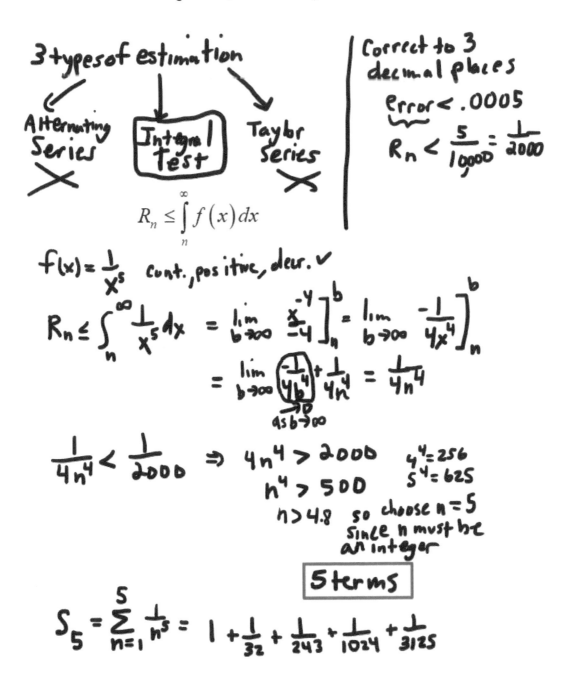

3 types of estimation

Alternating Series ✗

Integral Test

Taylor Series ✗

Correct to 3 decimal places

error < .0005

$R_n < \dfrac{5}{10000} = \dfrac{1}{2000}$

$R_n \leq \int\limits_{n}^{\infty} f(x)\,dx$

$f(x) = \dfrac{1}{x^5}$ cont., positive, decr. ✓

$R_n \leq \int\limits_{n}^{\infty} \dfrac{1}{x^5}\,dx = \lim\limits_{b \to \infty} \dfrac{x^{-4}}{-4}\Big]_n^b = \lim\limits_{b \to \infty} \dfrac{-1}{4x^4}\Big]_n^b$

$= \lim\limits_{b \to \infty} \left(\dfrac{-1}{4b^4}\right) + \dfrac{1}{4n^4} = \dfrac{1}{4n^4}$

as $b \to \infty$

$\dfrac{1}{4n^4} < \dfrac{1}{2000} \Rightarrow 4n^4 > 2000$ $4^4 = 256$

$n^4 > 500$ $5^4 = 625$

$n > 4.8$ so choose $n = 5$
since n must be an integer

5 terms

$S_5 = \sum\limits_{n=1}^{5} \dfrac{1}{n^5} = 1 + \dfrac{1}{32} + \dfrac{1}{243} + \dfrac{1}{1024} + \dfrac{1}{3125}$

4.36 Approximate the sum of the series to two decimal place accuracy.

Is the approximation an underestimate or an overestimate?

Find an expression for the sum.

$$\sum_{n=0}^{\infty} \frac{(-1)^n}{(2n+1)!}$$

4.36 Approximate the sum of the series to two decimal place accuracy.
Is the approximation an underestimate or an overestimate?
Find an expression for the sum.

$$\sum_{n=0}^{\infty} \frac{(-1)^n}{(2n+1)!}$$

$$\sum_{n=0}^{\infty} \frac{(-1)^n}{(2n+1)!} = 1 - \frac{1}{3!} + \frac{1}{5!} - \frac{1}{7!} + \frac{1}{9!} - \frac{1}{11!} + \cdots = \boxed{\text{Sin } 1}$$

$$= 1 - \frac{1}{6} + \frac{1}{120} - \frac{1}{5040} + \cdots$$

$$> \frac{1}{200} \qquad < \frac{1}{200}$$

Two decimal place accuracy

Alternating Series

error < 0.005

$|R_n| < \frac{5}{1000} = \frac{1}{200}$

$|R_n| < b_{n+1}$ $\Rightarrow$ Go until you get a term
$\nearrow$ that is
1st omitted term $< \frac{1}{200}$
add up the previous terms

$S_3 = 1 - \frac{1}{6} + \frac{1}{120}$

$|R_3| < \frac{1}{5040} < \frac{1}{200}$

$= \frac{120 - 20 + 1}{120} = \boxed{\frac{101}{120}}$

The error has the same sign as the first omitted term

$b_4 < 0 \Rightarrow R_3 < 0$

$S = S_3 + R_3 \Rightarrow R_3 = S - S_3 < 0 \Rightarrow S < S_3 \Rightarrow \boxed{S_3 \text{ is an overestimate}}$

Topic	Problems
Absolute Convergence	4.20 II, 4.21 II, 4.23 I
Alt. Series Test Approx.	4.36
Alternating Series Test	4.13 II, 4.20 I, 4.21 I, 4.24, 4.25, 4.26
Arclength	1.12-1.15
Center of Mass	1.19-1.21
Conditional Convergence	4.20 I, 4.21 I
Cross-Sections	1.1, 1.8
Direct Comparison Test	4.15 II, 4.17, 4.20 I, 4.22 II, 4.23 I
Disk Method	1.2, 1.11, 3.5
Geometric Series	4.7, 4.9, 4.10, 4.12, 4.22 I, 4.34
Improper Integrals	3.3-3.8
Integral Test	4.15 I, 4.18 I
Integral Test Approx.	4.35
Integration By Parts	2.1-2.5
Interval of Convergence	4.24-4.26
Limit Comparison Test	4.18 II, 4.21 I, 4.25
Linear Diff. Eq.	3.13, 3.16, 3.17, 3.18, 3.21
Maclaurin Series	4.27-4.29, 4.32, 4.33, 4.34
Mean	3.10
Median	3.12
Newton's Law of Cooling	3.22
Partial Fraction Decomp.	2.18-2.21
Probability	3.9-3.12
P-series	4.16 II, 4.26
Ratio Test	4.14 II, 4.19 I, 4.20 II, 4.21 II, 4.24-4.26
Root Test	4.14 I, 4.16 I
Separable Diff. Eq.	2.15, 3.14, 3.15, 3.19, 3.22
Sequence Limit	4.1-4.5
Shell Method	1.3, 1.6, 1.7, 1.9, 1.10, 2.17
Simson's Rule	3.2
Surface Area	1.16-1.18
Tank Problem	3.20, 3.21
Taylor Series	4.27-4.34
Telescoping Series	4.6, 4.8, 4.11
Test for Divergence	4.13 I, 4.19 II, 4.23 II
Trapezoid Rule	3.1
Trig. Powers	2.6-2.11
Trig. Substitution	2.12-2.17
Washer Method	1.4, 1.5, 1.7, 1.10